[redacted]
PERSONAL COPY.

Dry Strength Additives

Contributors

Chester T. Beals, P. E., Retired, Penick & Ford Ltd., Vancouver, Washington

Robert W. Davison, Research Scientist, Hercules Inc., Wilmington, Delaware

Lawrence A. Gaspar, Product Development Supervisor, National Starch and Chemical Corporation, Bridgewater, New Jersey

Donald S. Greif, Marketing Manager, National Starch and Chemical Corporation, Bridgewater, New Jersey

Shuji Maruyama, General Manager, R&D Office, The Japan Carlit Company, Ltd., Gunma, Japan

Charles S. Maxwell, Retired, Union Carbide Corporation, Burdett, New York

Walter F. Reynolds, Senior Research Chemist, American Cyanamid Company, Stamford, Connecticut

Morioki Shibuya, General Manager, Research Laboratory, The Japan Carlit Company, Ltd., Gunma, Japan

Fritz MK Werdouschegg*, Principal Consultant, P-T-B Consultants, St. Paul, Minnesota; commissioned under contract with General Mills Chemicals, Inc. (now Henkel Corp.), Minneapolis, Minnesota

*Presently Senior Process Chemist, Champion Packaging, Champion International Corporation, St. Paul, Minnesota

Dry Strength Additives

A Project of the Papermaking Additives Committee

Edited by Walter F. Reynolds

TAPPI PRESS

Library of Congress Catalog Card Number 79-67261
International Standard Book Number 0-89852-044-4

Preface

Over twenty years ago, Dr. John W. Swanson[1] summarized the state of the art of beater adhesives and fiber bonding by pointing out that a great many difficulties and inconsistencies arise when beater adhesives are used and much research must be done before some of the great potential can be realized.

The fact that this situation was taken seriously by the starch and resin suppliers is attested to by the sophisticated products and application technology that are available to the present paper industry. If the present committment to dry strength additives research is fruitful, the paper industry can expect even more effective dry strength additives to solve the problems of the future. For example, ever increasing utilization of waste paper, groundwood, high yield pulps, and pulps made from short fibered wood is going to require better dry strength products for paper manufacture. Application technology, e.g., cataphoretic mobility techniques, now available to the industry has helped immeasurably to maximize the performance of present day products. Inconsistent results still plague the papermaker, and a much better understanding of the variables affecting dry strength additives is still needed.

This book on dry strength products for papermaking represents an attempt to provide the paper technologist with a comprehensive treatise on the subject. The subject matter is arranged in such a way that the reader is first introduced to the subject by a discussion of the factors involved in the development of the strength of paper. The remaining chapters are arranged in the chronological order in which the dry strength products have been used in the paper industry, starting with size press starch and finishing with acrylamide polymers. The section on dry strength additives as practiced in Japan is included to emphasize the extent to which dry strength additives can be utilized in a country with very limited wood resources.

The authors contributing to this book were chosen on the basis of their contributions and knowledge of the specific subject. Editing was limited mainly to the mechanical aspects of writing without any attempt to standardize the style of presentations or to modify the meaning of any statement made by an author of a chapter.

It is suprising that relatively little has been published on dry strength additives for paper, a subject so vital to the paper industry. The majority of the published material has been contributed by starch, gum, and acrylamide resin suppliers. No reference is made to the older literature on unmodified starch, since this use of starch is practically non-existent.

The contribution of Richard T. Holzmann to this book is thereby acknowledged. As the original editor and Task Group Chairman, he guided the selection of topics and recruited most of the authors. Because of the press of business, however, he was unable to continue in these jobs. Thus, it became necessary to choose a new editor to complete the task.

The valuable assistance of the reviewing committee is also acknowledged. This group consisted of James P. Bambacht, Peter Calott, Lock-Lim Chan, William H. Crosier, Jr., Barry W. Crouse, Philip Luner, and John W. Swanson.

[1]*Tappi*, 39(5): 257-67 (1956).

Contents

Contents

Illustrations

Number	Title	Page

Tables

1

Theory of Dry Strength Development

Robert W. Davison

INTRODUCTION

Cellulose-based paper is essentially a laminated structure composed of cellulose fibers and bonds between these fibers. Paper has also been described as a bonded fibrous network which may consist only of cellulose, air, and water (*1*). The air within and between fibers contributes nothing to the structure's strength properties. Water, in the form of moisture held by the fibers and bonds, often has considerable influence on the strength of paper, as shown by the effect of ambient relative humidity on test results (*2*). With this exception, however, the dry strength of a paper structure depends only upon its fibers, fiber-fiber bonds, and the distribution of these two elements.

Deducing or explaining the mechanical properties of a multi-ply fibrous web which contains so many individual structural elements is quite a challenge, and paper scientists have applied themselves vigorously to this task for many years. The object of this chapter is to describe the primary factors which are now believed to contribute to the development of dry strength in paper. Those concepts which are well established are emphasized and documented. Where uncertainties still remain, the various current viewpoints are indicated briefly. The effects of practical processes (e.g., beating) are explained to the extent possible. The strength effects, both beneficial and adverse, of various types of chemical additives are mentioned briefly with only tentative explanations. This chapter is not concerned with explaining the mechanisms by which specific dry strength additives improve the strength of paper, although a few general comments are offered.

Structure of Paper

Paper is often viewed as an essentially two-dimensional structure because its surface dimensions are generally so large in comparison with its third dimension, thickness. Nevertheless, this third dimension is highly significant, since it represents the primary domain of the bonds which develop between fibers. A very thorough study has shown that paper is essentially a layered structure, with the fibers in a given layer lying primarily within that layer and oriented more or less randomly. Very little interweaving was observed between adjacent layers (*3*). From personal observation, a paper sheet having a basis weight of 60 g/m^2 (about 36 lb/3000 ft^2 ream) consists of perhaps 10 layers of fibers. Fiber-fiber bonding in the transverse or thickness direction is the primary link which holds these fibers together in the form of a tightly consolidated web or structure. A typical apparent density for this sheet is perhaps 0.6-0.7 g/cm^3, approximately half the density of its paper fibers. Thus, the volume of this fibrous web contains about 50% fibers and 50% air; most of the latter appears located between the fibers within each layer, as indicated by light micrographs of paper cross-sections (*5-6*). The thickness of this sheet is only 70-100 microns (7-10 $\times$ 10^{-3} cm), and the thickness of its *10* layers of fibers accounts for most of this third dimension.

The usual fibrous element in paper is wood pulp fibers, which have a length of 0.1-0.4 cm and are perhaps 20-40 micrometers wide and 5-10 micrometers thick. Fiber weight per unit length is often 3×10^{-6} g/cm (*1,7*); thus, 1 cm^2 of the sheet described above may contain 10,000 fibers averaging 0.2 cm in length. Since the average fiber is crossed, over and under, by 20-40 other fibers per 0.1 cm of fiber length (*8*), this 1 cm^2 piece of paper sheet may also contain several hundred thousand separate regions of fiber-fiber bonding. In addition to the very numerous fibers and the huge number of bonds lying within this small piece of paper, one must also consider the three-dimensional distribution of these structural elements, because structural uniformity (sheet formation) also has an important bearing on the dry strength of paper (*9-11*).

Dry Strength Properties

The dry strength of paper means different things to different people, depending upon the primary use for which the product is intended. Thus, the ability of paper to resist bursting stresses (its burst strength) may be the main consideration when discussing the structural properties required for a paper bag. In other applications, paper's resistance to rupture by tensile stress or its ability to resist rupture during repeated folding may be the primary requirement.

The common strength properties of paper are listed in Table 1.1. Most of these strength properties are increased by additional bonding between the fibers in a sheet. Transverse tensile strength and resistance to delamination are particularly dependent upon the degree of interfiber bonding. Sheet stiffness is generally increased by additional interfiber bonding, unless the latter greatly reduces sheet thickness. Since surface strength properties, such as abrasion resistance and pick resistance, require strong bonding of the surface fibers to those in adjacent underlying fibers, additional fiber-fiber bonding throughout the sheet tends to increase these properties also. However, a few strength properties (e.g., tear resistance) are generally impaired by increased fiber-fiber bonding. This conflict between the two groups of strength properties in Table 1.1 must be resolved by the papermaker to achieve an optimum balance, i.e., the best compromise in overall sheet strength characteristics for the particular intended end use. Analogous problems occur in the design of other practical structures (e.g., plastics, elastomers, and textile fabrics), where a compromise between tensile strength and structural flexibility is frequently required.

Table 1.1
Strength Properties of Paper

Increased by Additional Fiber-Fiber Bonding
In-plane tensile strength (breaking stress, breaking length)
Toughness (stress-strain work to rupture)
Mullen burst strength
Folding endurance
Transverse tensile strength (internal bond strength)
Stiffness (at constant sheet thickness)
Surface abrasion resistance
Surface pick resistance
Generally Reduced by Additional Fiber-Fiber Bonding
Elmendorf tear resistance
Compressibility
Bending flexibility (at constant sheet thickness)

Paper also has a number of non-strength physical properties, which are listed in Table 1.2. Here again, some of these are usually increased by additional fiber-fiber bonding, while others are reduced. One of the most important of these non-strength properties is sheet density, which represents the degree of web consolidation and thus indicates the extent to which adjacent fibers are in intimate surface contact. Because the increase in fiber-fiber bonding associated with beating increases sheet density, the latter has frequently been related to those strength properties which are improved by increased bonding and sheet consolidation (*12-13*).

Table 1.2
Non-Strength Physical Properties of Paper

Increased by Additional Fiber-Fiber Bonding
Apparent density
Air resistance (inverse of porosity)
Surface smoothness
Generally Reduced by Additional Fiber-Fiber Bonding
Sheet thickness (caliper)
Opacity
Surface softness
Compressibility

Since most of the strength properties of paper are increased by the same process (i.e., additional bonding between fibers), it seems reasonable to choose in-plane tensile strength as a representative property when describing the development of dry strength in paper. Accordingly, this chapter is concerned primarily with how and why tensile strength develops during the papermaking process, although other strength properties are mentioned where especially pertinent. Tensile strength is simply the ability of paper to resist being ruptured when one pulls on it. This property is often expressed quantitatively as the force required to tensile-rupture a paper strip of unit width (e.g., lb/in or kg/cm) when it is stressed in a planar direction. Alternatively, the breaking stress of a strip of paper is often reported per unit of sheet cross-section (e.g., kg/mm^2). Finally, the tensile strength may be normalized and reported as breaking length, generally in km. Breaking length refers to the length of paper having sufficient weight when hanging vertically to cause tensile rupture of itself and is, in principle, independent of the sheet basis weight. When consistent units are employed, breaking length is merely tensile strength per unit width, divided by sheet basis weight. In this chapter, the term tensile strength is used to include all of these expressions which describe the ability of paper to resist rupture when stressed within the plane of the sheet.

STRENGTH DEVELOPMENT MECHANISMS

To the neophyte, it may seem remarkable that a suspension of wood pulp fibers in water, screened to form a wet mat, can then be de-watered by three simple processes—drainage, wet pressing, and evaporation—to form a fibrous structure having considerable strength. An extensive research effort has been devoted to seeking and understanding the mechanisms by which this dry strength is developed. Some answers have been obtained, but many questions still remain.

Fiber-Fiber Bonding Concepts

It should be self-evident that the bonding between fibers is the very heart of dry strength development. A mass of short (0.1-0.4 cm long) fibers which remains unbonded does not produce a sheet of paper having either structural integrity or strength. Given a reasonably strong fiber furnish, one must develop at least moderate bonding between these fibers to produce a reasonably strong piece of paper.

To explain and control the development of dry strength in paper, one needs to know both the nature and the origin of fiber-fiber bonds. The following types of bonding are at least theoretically possible, and most have been proposed at one time or another:

1. Mechanical entanglement of fibers or surface fibrils,
2. Covalent bonding,
3. Ionic bonding,
4. Hydrogen bonding, and
5. Bonding by van der Waals forces.

Which of these is the primary type of bonding that develops between fibers during the papermaking process has been the subject of much study and debate during the last 50 years.

Some paper scientists have held the view that mechanical entanglement accounts for most or all of the fiber-fiber bonding in a dry sheet of paper (*14*, *15*). The interweaving of the fibers within a complex fibrous mat would certainly be adequate to account for considerable dry strength development, but a recent study has demonstrated the virtual absence of this process (*3*). Paper is not basically an interwoven fibrous mat. Alternatively, mechanical entanglement could involve the many fibrils or hairs which can be seen, using the light microscope, on the surfaces of beaten pulp fibers (*14*). A homely analogy might involve the sticking tendency of two hair brushes whose many bristles are intimately mingled. However, this viewpoint is not accepted by most research workers. The electron micrography evidence of Jayme and Hunger (*5*) clearly demonstrates strands consisting of microfibril bundles and remnants of beaten fiber outer walls spanning between adjacent fibers in paper, but this type of fiber-fiber bonding does not represent mechanical entanglement.

When untreated (waterleaf) paper is immersed in water, it loses essentially all of its strength (i.e., it has no wet strength). Covalent bonding (50-100 kcal/mole), second in the above list of possibilities, has been proposed in attempts to explain the wet strength which is imparted to paper by the addition of various wet strength additives (*16*). Because this wet strength improvement sometimes tends to be accompanied by an equal increase in dry strength, it has been suggested that the increased dry strength is also the result of covalent fiber-resin-fiber bonding (*17,18*). However, covalent bonding as an appreciable source of bonding between fibers in an untreated sheet of paper has not been seriously advanced, because it would be difficult to explain why a covalently bonded structure so readily loses its dry strength by simple immersion in water at room temperature.

Ionic bonds (10-30 kcal/mole) could account for strong fiber-fiber bonding which would nevertheless be completely disrupted by water. Further, carboxyl groups in wood pulp fibers (1-10 meq/100 g) could conceivably lead to a considerable amount of ionic bonding if fiber surfaces contained a generous number of these carboxyl groups. With the assistance of metal ions in the aqueous phase, one can easily visualize the development of ionic bonds of the following type:

$$\text{fiber-COO}^- \text{— cationic metal ion —} {}^-\text{OOC-fiber}$$

Calcium or aluminum would be the most likely metals to participate in these ionic bonds. The concept that ionic bonding is the chief source of fiber-fiber bond development in untreated (waterleaf) paper has not been strongly urged by paper scientists, probably for lack of specific evidence. However, Allan and Reif (*19,20*) have recently reported work involving surface-modified fibers in which certain sheet strength improvements are cited to demonstrate the development of considerable ionic bonding between these specially treated fibers.

During the last 25 years, most paper scientists have accepted the concept that hydrogen bonding (4-6 kcal/mole) between the vast numbers of hydroxyl groups on the surfaces of adjacent cellulose fibers (and fibrils) in a fiber mat is the primary source of fiber-fiber bonding in dry paper. Many indirect experiments which strongly support the hydrogen bond viewpoint have been reported. Before summarizing this evidence, however, it is perhaps worthwhile to note the logic supporting the concept of hydrogen bonding between cellulose fibers. First, the potential bonding material, cellulose hydroxyl groups, is readily available on the surfaces of hydrated wood pulp fibers. In fact, it is almost universally agreed that the hydration (intimate association of water molecules with cellulose) of these fiber surfaces represents profuse hydrogen bonding between water molecules and surface hydroxyl groups. It seems rather simple for this hydrogen bonding to be transferred into hydrogen bonding between the hydroxyl groups of adjacent fiber surfaces as the latter are brought into very intimate contact during the simultaneous removal of the water molecules from a wet fiber mat. This suggested mechanism for the development of fiber-fiber bonding is illustrated in Figure 1.1. The top diagram represents a beaten fiber

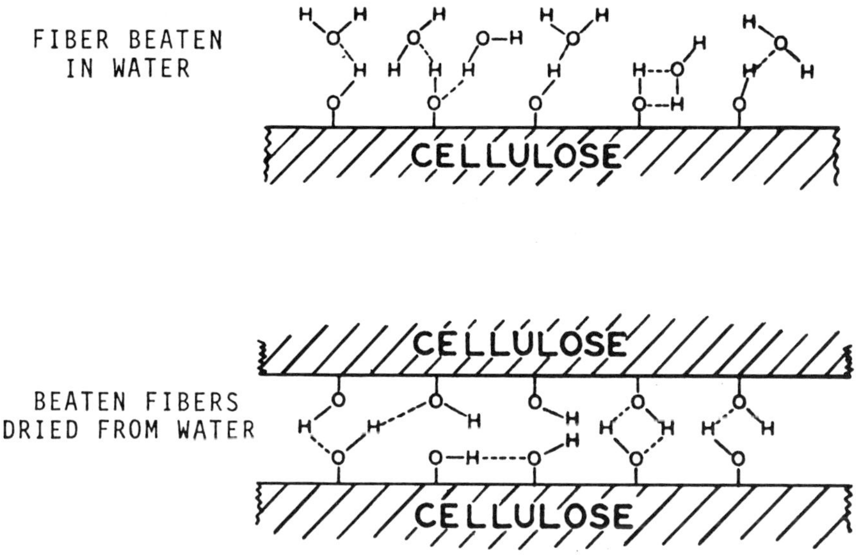

Fig. 1.1. Formation of hydrogen bonds between adjacent fibers.

surface which is strongly hydrated by the hydrogen bonding of water molecules. When a wet fiber mat is formed, pressed, and dried, adjacent fibers come into close contact with each other; and interfiber hydrogen bonding can develop between cellulose surfaces as the water molecules are removed. Thus, the bottom diagram in Figure 1.1 shows two adjacent fibers which have become hydrogen bonded while being dried. Second, it has been asserted for years that the primary source of the great strength of cellulose fibers themselves is massive hydrogen bonding between the hydroxyl groups of adjacent cellulose polymer chains, which are the basic structural element in the fiber's morphology (*21,22*). If hydrogen bonding accounts for the tremendous tensile strength of cellulose fibers (often approaching 100,000 psi), it would seem that this process would also provide the relatively limited bond strength between fibers in paper. Finally, there are many confirmed examples of the relative ease with which hydrogen bonds develop (e.g., water, alcohols). It would appear that hydrated cellulose surfaces should not be reluctant to participate in the same process.

The scientific evidence which has been amassed to support hydrogen bonds as the main source of bonding between fibers is quite convincing, even though it does not involve direct critical experiments. It has been demonstrated that wet pieces of smooth cellophane film become well bonded if they are dried while being held in close surface contact (*23*). This observation would seem to rule out the mechanical entanglement mechanism in favor of such a chemical bonding process as hydrogen bonding. Various publications have reported that the formation of a wet fiber mat from an aqueous slurry, followed by the solvent-exchange replacement of the water by a series of liquids of decreasing polarity and then by drying, produces a relatively weak sheet of paper (*24,25*). Thus, the presence of water or some other highly polar liquid capable of hydrogen bonding seems required for the development of good dry strength in paper. A similar experimental approach involved the freeze-drying of a fiber mat formed from an aqueous slurry, followed by sublimation to remove the water from the sheet (*25-27*). The resulting dry paper was very weak, presumably because the absence of liquid water during the de-watering of the wet mat prevented the development of interfiber hydrogen bonding. Using heavy water, Corte and Schaschek (*28*) carried out deuterium exchange experiments involving both isolated cellulose fibers and paper made from these fibers. Infrared analysis indicated that most of the hydroxyl groups of the isolated fibers were involved in intrafiber hydrogen bonding. However, their deuteration results demonstrated that up to 2% additional hydroxyl groups became bonded during the papermaking process. This experimental result is often cited as convincing proof of the existence of interfiber hydrogen bonding in paper (*1*).

Nevertheless, persistent skeptics feel that the case for hydrogen bonds as the main source of fiber-fiber bonding has yet to be proved. It can be argued that the imposition or assumption of hydrogen bonding is not necessary, because the ever-present van der Waals forces (0.5-2 kcal/mole) are sufficient to account for the bonding between adjacent fibers which are in very close large-area contact. Since any two surfaces stick together to some extent if they are in very close contact (i.e., approach within a few Angstrom units), it seems that this should be the case for paper fibers in a tightly consolidated dry web. The amount of close bonding contact between two adjacent fibers in a sheet of paper is another area of considerable scientific disagreement and is discussed later. However, a very weak type of bonding spread over a large area of contact should provide the required amount of bonding between two fibers just as easily as stronger bonds over a much smaller contact area. Thus, van der Waals forces could be sufficient, and the cited experimental evidence which suggests hydrogen bonding does not exclude the distinct possibility that the weaker van der Waals forces may account almost entirely for fiber-fiber bonding.

In summary, the specific nature of fiber-fiber bonding is still somewhat uncertain. Generally, mechanical entanglement, covalent bonding, and ionic bonds appear to play little or no part. Hydrogen bonding is a plausible and adequate mechanism. However, van der Waals forces offer another possibility which cannot be completely ignored. These latter forces could provide weak bonding between all parts of adjacent cellulose molecules, not merely between their polar hydroxyl groups.

Fiber-Fiber Bond Strength

The inherent bond strength per unit bonded area and the extent of bonding (i.e., the true bonded area involved in a fiber-fiber crossover region) are so intimately involved as a product factor that an almost infinite number of combinations can produce the same observed level of fiber-fiber bond strength. Unfortunately, the true bonded area between two fibers in close contact is another undecided question. As shown by Figure 1.2, there is considerable doubt as to whether complete bonding exists in the crossover area (about 2000 square microns) between two fibers. The strong effort made by Page and co-workers (*8,29-31*) revealed that the optical contact area between two bonded fibers ranges from 50 to as much as 1000 square microns or more. The area of optical contact represents surfaces too close together to refract or scatter incident light—that is, a region where the two fibers are separated by less than about 150 Angstrom units. Much of this optical contact area may not involve bonding, however, because two cellulose surfaces must approach within three Angstrom units for interfiber hydrogen bonding to develop. An approach within about five Angstrom units or less would be needed to develop appreciable bonding by van der Waals forces. As is suggested in Figure 1.2, perhaps only the central part of the fiber crossover area is actually bonded.

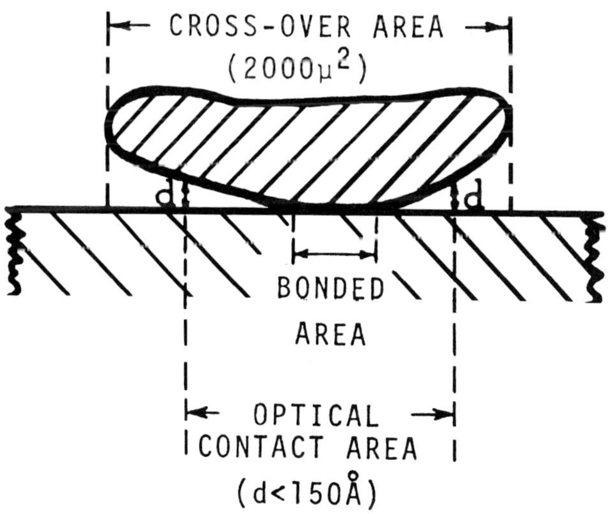

Fig. 1.2. Areas of interest in fiber-fiber bonding.

Efforts to measure bonded area by nitrogen adsorption (*32,33*) indicate that bonded area changes measured by this technique are reasonably proportional to changes in optical contact area. Since only spaces of less than four Angstrom units between fibers should restrict the adsorption of nitrogen gas molecules, these results suggest that most of the optical contact area is really bonded. On the other hand, Jayme and Hunger (*5*) have reported rather convincing electron micrography evidence suggesting that only 10-20% of the optical contact area represents true bonding. They noted that it is not the surfaces which come into contact but microfibrils attached to these surfaces, as is shown by electron microscopy. Proceeding even further, Nissan and Sternstein (*34*) have invoked the "invariance of cellulosic fibrillar patterns with magnification" to argue speculatively that further magnification to the atomic level would reveal that less than 1% of the optical contact area is truly bonded.

The shear strength of individual fiber-fiber bonds has been determined by various groups of research workers to be 0.4-1.6 g (*7,35,36*). This finding is supported by others who have measured the bond strength between a single cellulose fiber and a cellulose shive (*37*) or between a single fiber and cellophane film (*37,38*). When the results are normalized for apparent contact area, they support the value range cited above for individual fiber-fiber bonds. This experimentally determined bond strength is much less than the value of almost 100 g, which was calculated by Davison on the assumption that the entire crossover area between two fibers is hydrogen bonded (*7*). He suggests that the big discrepancy between calculation and experiment may indicate that only a very small fraction of the optical contact area shown in Figure 1.2 represents actual bonding between fibers. The discrepancy noted would, of course, become several-fold less if bonding by van der Waals forces, rather than hydrogen bonding, were assumed.

In summary, the fraction of crossover area between two fibers which is actually bonded is not known; current opinions range from less than 1% to as high as 50% or more. Because neither the nature of fiber-fiber bonding nor its extent per crossover region is known at present, one must proceed only with the empirical knowledge that the fibers are held together within a sheet of paper by individual fiber-fiber bonds which break under an applied load of less than 2 g.

Strength Development in the Wet Mat

Regardless of the nature and extent of the bonds between fibers in paper, there is intense practical interest in the mechanism by which these bonds develop during the papermaking process. An early insight was provided by Campbell (*39*), who stated that various surface tension effects during the removal of water from a wet fiber mat ultimately provide a powerful internal force which pulls adjacent fiber surfaces into very intimate and extended contact, thus producing strong fiber-fiber bonds. As shown by Figure 1.3, this cal-

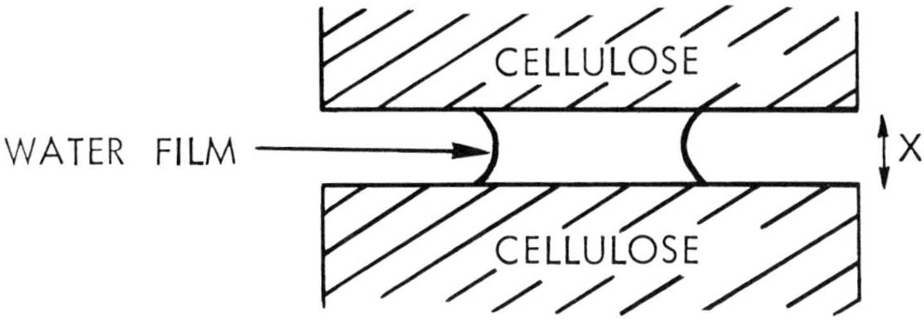

$$\text{TENSION} = \frac{2\gamma}{X} \text{ DYNES/CM}^2 \text{ OF WATER FILM}$$

$$\text{FOR } \gamma = 72 \text{ DYNES/CM AND } X = 72\,\text{Å}$$

$$\text{TENSION} = 2 \times 10^8 \text{ DYNES/CM}^2 \approx \underline{3000 \text{ psi}}$$

Fig. 1.3. Internal tension exerted between adjacent wet fibers because of capillary action.

culated internal tension amounts to 3000 psi for two wetted areas which are only 72 Angstrom units apart. Campbell's thesis has a sound theoretical basis. Although this strongest internal tension effect has not been demonstrated directly with regard to the bonding of paper fibers, no one has refuted his concept; and the "Campbell effect" is frequently referred to and accepted by modern paper scientists.

Interfiber relationships develop even in rather dilute pulp slurries. The fibers tend to form a continuous network, as demonstrated by the elastoviscometer measurements of Steenberg and co-workers (*40*). These interactions increase with increasing slurry consistency, and the initial strength of a wet fiber web merely represents similar fiber interactions which have developed to a much higher level of intensity. The development of wet web strength with increasing dryness is shown in Figure 1.4, which represents the findings of Lyne and Gallay (*41*). For wet webs below 10-12% solids, web strength is quite low and is due entirely to the frictional resistance which develops between fibers when the web is stressed. At about 12% wet web solids, air starts to enter the fiber-water structure, and web strength increases sharply, due to the surface tension effect at the fiber-water-air interface. At about 28% solids, all of the interfiber water has been replaced by air, except at fiber crossover regions; this effect of surface tension on web strength therefore passes through a maximum value. According to Lyne and Gallay, interfiber bonding starts to develop vigorously at this stage of dryness, so the web strength curve in Figure 1.4 passes through an in-

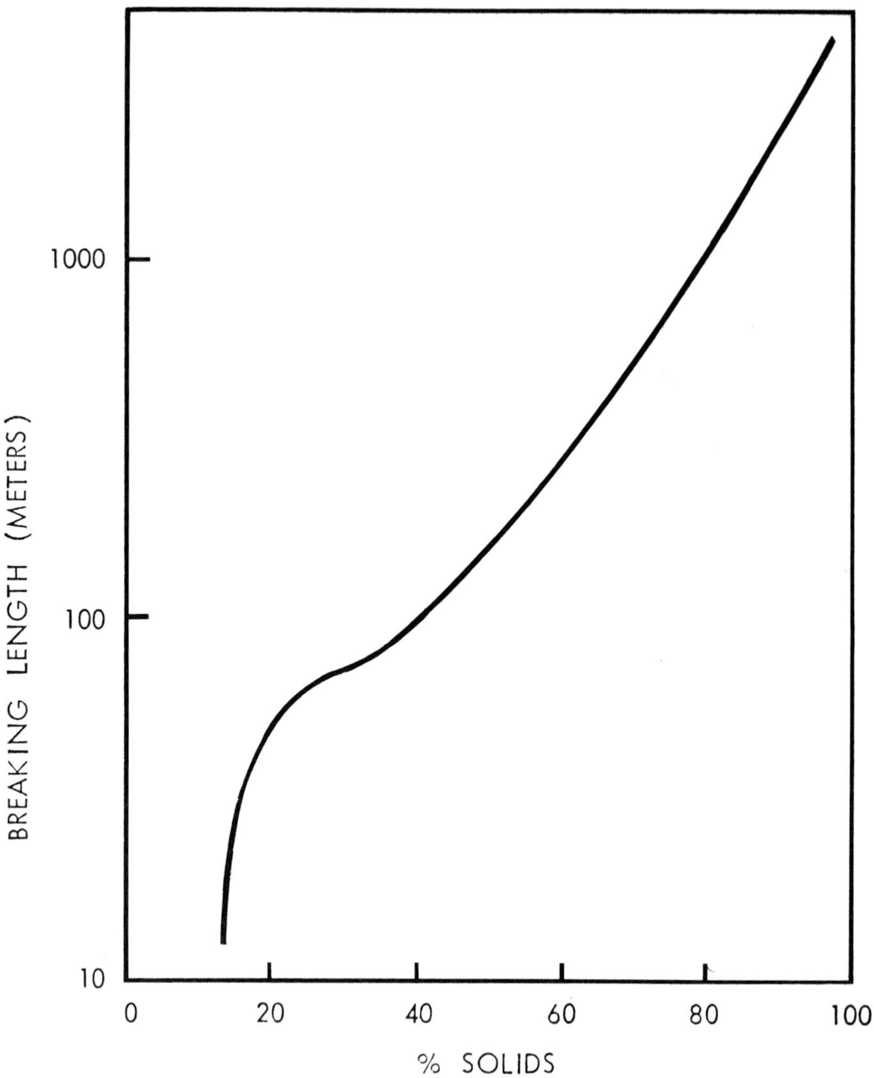

Fig. 1.4. Development of web strength with increasing dryness (*41*).

flection point and then increases even more sharply, all the way to essentially complete dryness of the sheet. To a large extent, the Lyne and Gallay explanation of strength development in a wet fiber web is supported by Robertson's more recent work (*42*). Robertson described a second point of inflection, however, which he assigned to the onset of hydrogen bond development between adjacent fibers at about 50% web solids. This may also be the point at which

Campbell's final and most powerful internal tension effect (see Figure 1.3) starts to manifest itself.

In addition to summarizing the strength development stages described above, various other events which take place during the removal of water from a wet fiber web are listed in Figure 1.5. Thus, optical contact areas are reported by Page and Tydeman (*43*) to make their initial appearance at about 45% web solids, close to Robertson's second inflection point. Water-swollen fiber walls have been described by Stone and Scallan (*44*) as consisting of 100-200 lamellae or slit pores. They reported that the volume of these slit pores starts to decrease at about 50% web solids. Finally, the transverse shrinkage of drying fibers begins at some stage of dryness well above 45% web solids, possibly as high as 65% solids (*43*). The panorama of events summarized in Figure 1.5 suggests that the plasticity and deformability of water-swollen fibers, encouraged by Campbell's most powerful surface tension effect, may greatly increase the fiber-fiber contact area during the de-watering of a wet fiber web.

In summary, Campbell's surface tension effects provide adequate forces for the strong encouragement of intimate contact and bonding between adjacent fiber surfaces during the sheet de-watering process. Thus, wet fiber deformability may contribute much more to fiber-fiber bonding and the development of strength in the drying web than is contributed by the inherent adhesive or bonding tendency of the fiber surfaces. Extensive fiber-fiber contact resulting from wet fiber deformability is sometimes referred to as the "boiled macaroni" theory of dry strength development.

Sheet Strength Limitations

Although paper is remarkably strong, its strength properties are limited by the strength of its structural elements (fibers and fiber-fiber bonds) and by several other factors (e.g., fiber length, sheet formation, local weak spots, or flaws in the sheet).

In principle, the ultimate strength of a very strongly bonded fibrous structure should depend solely on the fiber strength. Single fiber strength testing has revealed that pulp fibers produced from a variety of wood species break under axial tensile loads ranging from 5 g to 50 g, more generally 10-30 g (*7,37,45-49*). The calculated tensile strength of a well bonded sheet composed of these fibers agrees fairly well with zero-span tensile measurements of paper strip specimens (*45*), but both values are frequently two to three times greater than the normal-span tensile strength of this paper (*7,45*). This discrepancy demonstrates that factors other than axial fiber tensile strength often tend to limit the strength of paper. Careful study by electron microscopy of the ruptured zones of very strong, well bonded papers has suggested that transverse rupture of fiber walls may have been the strength limiting factor for these specimens (*50*).

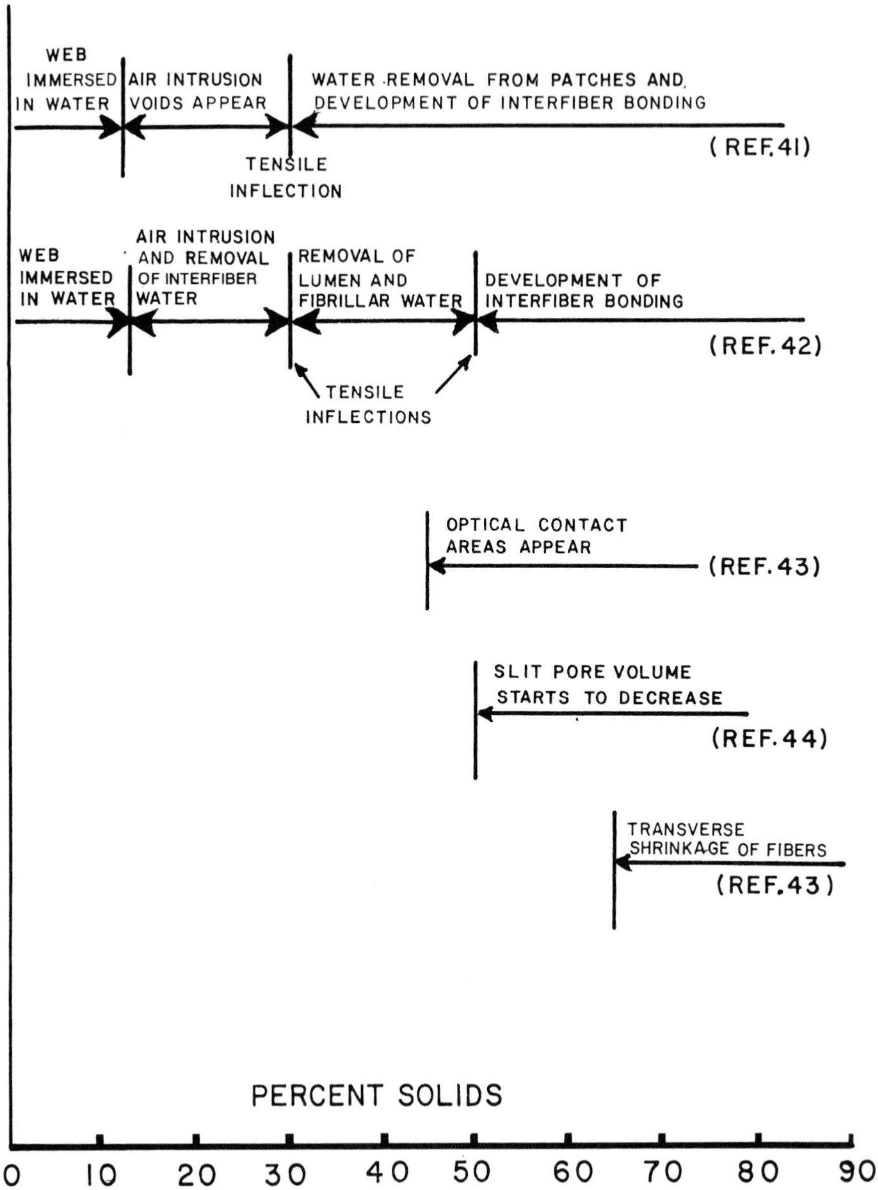

Fig. 1.5. Events which occur during removal of water from wet fiber web.

The tensile strength of paper can be greatly increased by various means (e.g., beating, wet pressing, dry strength additives) which do not increase single fiber strength enough to explain this strength improvement. It is generally accepted that these strengthening effects are due, instead, to improved fiber-fiber bonding. As mentioned earlier, the measured shear strength of individual fiber-fiber bonds (i.e., the bond between two fibers in a single crossover region) is 0.4-1.6 g (*7,35,36*). Therefore, the bonds between fibers appear to be much weaker than the fibers themselves. How these weak bonds can allow paper to make full use of its strong fibers is suspect. It is possible that the shear strength measurement of individual bonds actually involved a peeling rupture process and thus did not yield the true bond strength; however, the results reported by various groups of research workers agree quite well and were obtained by techniques designed to avoid the peeling process. Alternatively, the fiber-fiber bonds in a well consolidated sheet may tend to act in concert, rather than singly. In this case, the cooperative restraint imposed by perhaps 10 or 15 of the many bonds connected to a single fiber in the sheet would be sufficient to cause fiber rupture during tensile testing of the sheet, as suggested by Page (*51*). On the other hand, a variety of qualitative evidence has been reported which suggests that fiber-fiber bonds break singly and consecutively during the tensile stressing of paper (*7,35,52*).

During most of the last 50 years, paper scientists have generally agreed that fiber-fiber bond strength is the factor limiting the dry strength of commercial paper. Recently, however, the work of several investigators (*45,51,53,54*) has strongly suggested that the fibers themselves limit the strength of extensively bonded paper. A lively discussion has followed. Accumulated evidence to support the viewpoint that fiber-fiber bonds are the weak link in paper dry strength has been summarized by Davison (*7*). However, a theoretical strength equation introduced by Page (*51*) states that the bonds will break in weakly bonded paper but that fibers must first rupture if the paper is highly bonded. This equation is based on the assumption that the bonds attached to a given fiber act in concert, not singly, to resist an applied stress. The Page equation is supported by certain experimental evidence (*51*) and has been accepted as a good working hypothesis in several very recent studies (*55,56*). Other recent research has employed the approach that local stress intensification during paper tensile testing limits sheet strength to a level considerably lower than that which can be calculated from single fiber strength values. The relative importance of fibers vs. fiber-fiber bonds in limiting sheet strength remains uncertain, may be simply a matter of degree, and probably depends on the specific sample being studied.

The effect of fiber length on sheet strength properties has been studied intensively during the last 50 years. Early findings by Clark (*57*) suggested that paper tensile strength depends on the square root of the "weighted average fiber length by weight" (L), burst strength on the first power of L, and tear resistance on L raised to the 3/2 power. More recent studies by others have concluded that strength properties are much less dependent upon fiber length (*4,58*). It is generally felt that tear resistance, folding endurance, and sheet rigidity or stiffness are the only strength properties which are greatly affected by fiber length, but the issue still seems cloudy.

Nonuniform distribution of the fibers and fiber-fiber bonds within paper (poor sheet formation) can limit sheet strength properties by producing obvious weak spots or flaws. In general, poor sheet formation impairs the tensile and burst strength values but increases tear resistance (*9-11*). In a more general sense, any sort of local imperfections or flaws in a sheet of paper can lead to inefficient use of the fibers and fiber-fiber bonds. It does not matter whether these weak spots are present initially (poor formation), are produced during test preparation (ragged cut edges of strips), or develop randomly during the stressing process. The effect of these local weak spots is demonstrated by the inverse dependence of paper tensile strength on the length of sample tested (*7,54*). It is generally agreed that the presence of tiny weak spots or flaws in a test specimen may lead to severe localized overloading during the stressing process. Local stress intensification causes the development of incipient fracture zones and eventually the propagation of fracture to produce rupture of the paper (*59-62*). The severe limitation of paper's resistance to tearing by the degree of fiber-fiber bonding, which localizes and thus intensifies the applied tearing stress, is a somewhat analogous effect which will be discussed later.

Sheet Strength Analysis and Description

Instead of approaching the strength of paper by considering the strength of its structural elements, one can examine the various properties of the sheet and then deduce their origins. A very elegant attempt to develop a purely scientific hypothesis to explain the strength of paper was contributed by Nissan (*63,64*). Assuming that fiber-fiber bonds are due to hydrogen bonding between the hydroxyl groups of adjacent cellulose fibers, Nissan used a thermodynamics approach to develop a one-dimensional hydrogen bond theory to describe paper dry strength. His concept was later extended to develop two- and three-dimensional theories for hydrogen-bonded networks (*65*). Many other scientists have employed mathematical-mechanical models of paper structure to explain sheet strength properties, especially tensile stress-strain relationships, as summarized in several recent review articles (*1,66,67*). Some of these models appear to describe adequately the initial linear slope (elastic region) which is observed in the typical load-elongation (stress-strain) curve for paper (the region up to the yield point in Figure 1.6). However, none of these models has explained satisfactorily the plastic flow or creeping tendency represented by the

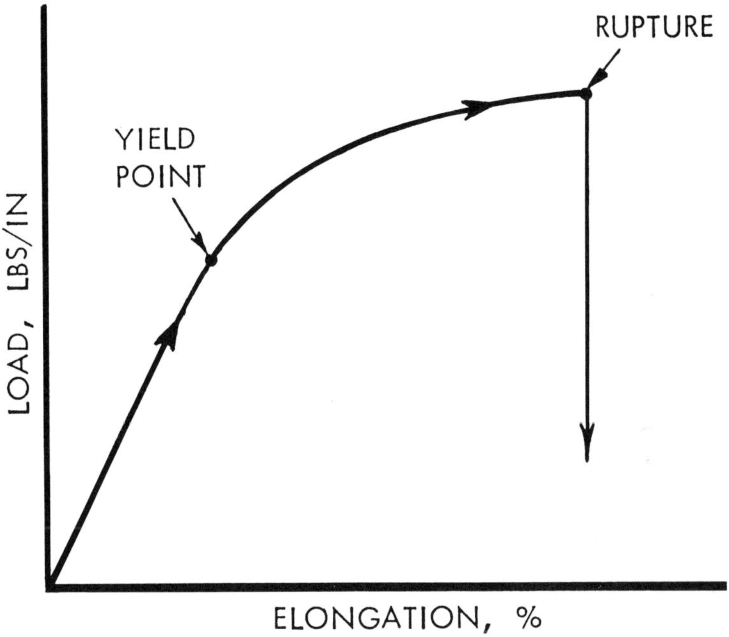

Fig. 1.6. Typical load-elongation curve for dry paper.

curvature from the yield point to the point of paper rupture in Figure 1.6. Structural models have not been developed to account satisfactorily for the other more complicated strength properties of paper (e.g., burst, tear). The large number of fibers and huge number of fiber-fiber bonds which make up even a tiny piece of paper pose an almost insurmountable challenge to model builders, even with the assistance of computers.

Several empirical relationships have been developed to describe the strength properties of paper. One early effort involved the dependence of various strength properties on the sheet density (*12,13*). This approach has been quite popular, partly because the parameters involved are relatively easy to measure and partly because sheet density seems to reflect the degree of fiber-fiber bonding quite reliably. Increased bonding between fibers, in turn, greatly improves tensile, burst, fold, and several other strength properties, while generally reducing tear resistance. These empirical sheet density relationships still find use in various practical situations.

Empirical relationships have also been established between the various sheet strength properties themselves. Thus, tensile strength and burst strength are usually found to increase and decrease together, even for papers made under various circumstances. The inverse relation between tear resistance or sheet flexibility on the one hand and tensile or burst strength on the other has also attracted some attention (*68-70*), since these two groups of properties generally respond in opposite fashion to an increase in fiber-fiber bonding. More complicated functions involving an empirical relationship between two or more paper strength properties have been described (*71,72*) but appear to have limited utility.

In summary, empirical working relationships which help to describe many variations in paper strength properties have been devised, but it seems unlikely that a theoretical model which explains the origin and limitations of these properties will be developed soon because of the extremely complicated nature of this fibrous structure.

PRACTICAL EFFECTS ON PAPER DRY STRENGTH

To be useful, as well as credible, the mechanisms by which paper develops dry strength should explain the many practical effects, both physical and chemical, which have been found to affect sheet strength. These effects include those due to the fiber furnish, its processing, the various steps in the papermaking process, effects of environment, and the effects of dry strength agents and other paper chemical additives. Concerning this vast array of strength parameters, the published literature contains a huge number of highly specific experimental observations but only a limited number of fundamental explanations. To the extent possible, simple explanations of these strength effects based on the various mechanisms described earlier are suggested.

Effects of Fiber Components

The various types of wood pulp fibers used in papermaking affect sheet strength for many different reasons. General fiber strength and length effects have already been discussed. Very recent elaborate experiments have demonstrated rather conclusively that the strength of single fibers depends considerably on the angle of fibril winding relative to the fiber axis (*73*). Fiber strength also seems to depend on the absence of weak spots or flaws in the fiber wall. The presence of sufficient hemicellulose to provide adhesive for the various structural components of the fiber wall is also believed to contribute to single fiber strength (*37,47*).

The ease of fibrillation and hydration of pulp has a tremendous effect on the dry strength of the paper product and thus leads to the concept that "paper is made in the beater." It is generally agreed that high yield kraft, waste or recycled paper, and hardwoods are "weak" pulps; i.e., these fibers produce relatively weak paper. The high lignin content of high yield kraft (*74*), the hornified regions of recycled pulp (*5,75*), and the relatively thick fiber walls and shorter lengths of hardwood fibers (*4*) are the current explanations of these effects. These "weak" pulps are all relatively difficult to hydrate and plasticize, although for different reasons. A high lignin content makes pulp fibers more hydrophobic than cellulose; stiff hornified (welded) clusters of cellulose molecules or microfibrils are more difficult to hydrate; and thick-walled fibers are less likely to form collapsed ribbons having high contact area, even when well hydrated and swollen. However, fibers which are easily hydrated and plasticized are more inclined to form extensive, strong fiber-fiber bonds during the papermaking process, in accordance with the "boiled macaroni" theory.

Various techniques have been employed to modify cellulose fibers so that they will form stronger paper. A moderate introduction of acetate ester groups was found to improve the ability of pulp fibers to make strong paper, although excessive substitution of this relatively hydrophobic group impaired strength (*76-79*). The slight substitution of either carboxymethyl (*80,81*) or hydroxyethyl (*82-84*) groups in hard-to-beat rag or cotton linter fibers greatly improved the ease of hydration of this furnish. More recently, several investigators have reported improved cellulose-to-cellulose water-induced bonding after the brief corona discharge irradiation of cellophane or paper surfaces (*85, 86*). Irradiation of synthetic plastic sheets has also improved their adhesion to cellulose (*87*). Probably all of these surface-modification treatments produce the same effect, i.e., a relatively disordered surface on a molecular scale, thus leading to improved bonding to adjacent "activated" surfaces.

With regard to surface adhesion, wood fiber components, such as lignin and hemicelluloses, can be dealt with in the same manner. The hydrophobicity of lignin should certainly reduce bonding between adjacent fiber surfaces in an aqueous system, while the very hydrophilic water-swellable hemicelluloses should enhance fiber-fiber bonding. Hemicelluloses located inside the fiber walls should also aid in the hydrating and swelling (hence plasticizing) of these walls to provide more conformable fibers for eventual strong interfiber bonding. Accordingly, the addition and removal of various types of hemicellulosic materials has been shown to have considerable effect on the dry strength of paper (*74,88,89*). It has been proposed that hemicelluloses leached from pulp fibers by water may also provide a natural adhesive for fiber-fiber bonding by adsorbing or precipitating back onto fiber surfaces during the de-watering of paper (*90,91*). The presence of pulp fines in a wet fiber mat has also been reported to improve the strength of the dried structure (*92*), but this effect is not altogether accepted at present.

To summarize, the natural presence in or addition to cellulose fibers of hydrophilic polymeric materials improves surface adhesion and/or internal plasticization of fibers, thus improving fiber-fiber bonding. Conversely, the presence of hydrophobic material impairs bonding. Processes which "activate" fiber surfaces also improve bonding.

Processing Effects

Based on the previous section, chemical pulping and bleaching processes should affect the strength of paper by the extent to which they extract hydrophobic materials (lignin, resins, wax) while leaving the hydrophilic hemicelluloses in the treated fibers. In general, this rule is supported by practical experience. Unfortunately, however, most pulping and bleaching processes are nonspecific and tend to remove both the hydrophobic and the hydrophilic impurities in cellulose fibers, and thus a compromise is generally adopted in practical papermaking systems. Groundwood, the product of purely mechanical pulping, is an extreme example of a truly weak fiber furnish being acceptable in exchange for a very high yield for certain applications in which high sheet strength and durability are not required.

The beating or refining of pulp is a very complicated physical process by which many of the properties of pulp fibers are drastically altered, as reflected by the effects of beating on paper dry strength. Beating consists of the mechanical abrasion of pulp fibers in the presence of a very polar penetrating liquid, such as water, in order to produce a high degree of swelling by hydration of the cellulose molecules composing the fibers. Thus, experiments in which pulp fibers were beaten in a nonpolar fluid produced extremely weak paper from a subsequent aqueous fiber slurry, while beating in formamide, a highly polar liquid, produced normal strength paper from an aqueous slurry (*93*). Among the many publications concerning the beating process, a comprehensive review article by Higgins and De Yong (*92*) describes how the effects on fibers affect sheet strength properties. Beating produces the following effects:

1. Cutting, shortening, and weakening of fibers;
2. Production of free cellulose fines;
3. Solubilization of hemicelluloses;
4. Surface fibrillation, hydration, and gelatinization; and
5. Internal fibrillation, hydration, and plasticization.

The first of these refining effects improves sheet formation considerably, thus contributing to paper dry strength in this way. However, the effects of fiber shortening and weakening can affect sheet strength adversely, as discussed earlier. Production of fines greatly reduces slurry freeness or drainage rate, moderately improves sheet formation, and improves certain sheet strength properties by providing more fiber-fiber contact area during papermaking. To the extent that solubilization of hemicelluloses provides useful adhesive for fiber-fiber bonding, the sheet strength properties are affected (*91*). The last two effects listed above have the greatest impact on paper dry strength. Surface fibrillation refers to disrupting the surface elements of a wood pulp fiber. It can be thought of as a scuffing process applied to organized bundles of cellulose fibrils located on or near the fiber surface, which partially frees the individual fibrils of these bundles. Surface fibrillation of pulp fibers in water leads to the swelling (i.e., hydration) of these fibrils, increases fiber surface area, and makes these surfaces somewhat tacky and adhesive. Internal fibrillation can be regarded as a similar disorganizing process applied to bundles of fibrils located in the interior of the fiber wall. This mashing or crushing process, when carried out in the presence of a swelling agent such as water, leads to considerable swelling and separation of the individual interior fibrils (i.e., hydration). Internal fibrillation results in a plasticized deformable fiber and improves fiber-fiber contact area when a wet fiber mat is pressed and dewatered. Both of these beating effects contribute strongly to improved bonding between fibers and markedly improve most sheet strength properties (e.g., tensile, burst, fold) while impairing tear resistance.

The opposing effects of beating on sheet strength properties are summarized in Figure 1.7 by a typical "beating curve," in which Elmendorf tear resistance is plotted against either tensile or burst strength. The concept of tear-tensile and tear-burst strength plots was introduced some years ago as a means of representing the sheet strength balance or limitation imposed on these two groups of properties (*68*). Thus, curve ABCD in Figure 1.7 represents the complete beating curve for pulp. Point A represents handsheets made from defibered but unbeaten pulp. These bonded sheets have low tear resistance and are also weak

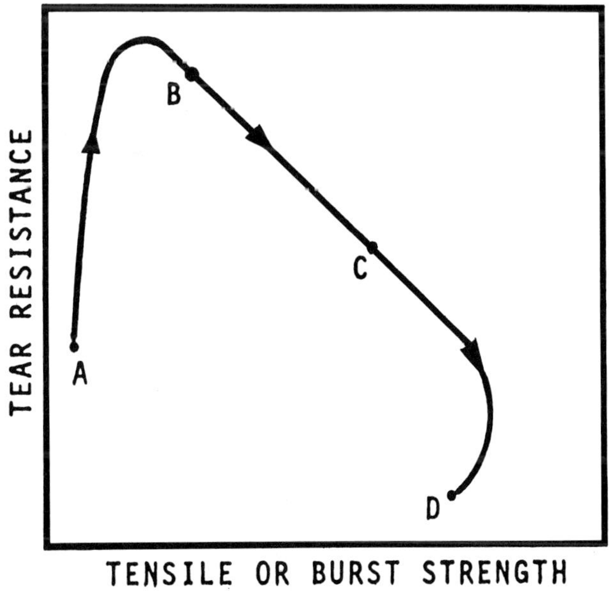

Fig. 1.7. Effects of beating on dry strength of paper.

in tensile and burst strength. As beating proceeds, both groups of properties increase initially, but tear resistance soon passes through a maximum and then decreases with further beating. However, the tensile and burst strengths continue to be improved, due to increased fiber-fiber bonding. Only if beating is carried to an extreme do the tensile and burst values finally start to decrease, probably because extreme refining has cut the fibers into short segments. Point B on this plot represents the strength of handsheets made from lightly beaten pulp; point C represents moderately-to-heavily beaten pulp; and point D represents pulp which has been overbeaten. The linear part of the curve represents the normal beating range.

The maximum in tear resistance is due to the fiber-fiber bonding which develops as beating proceeds. The tear resistance of paper made from unbeaten pulp is low because there is essentially no bonding to hold the fibers together. A slight amount of beating greatly improves tear resistance by providing some bonding between the fibers. However, as beating proceeds further, increased fiber-fiber bonding eventually tends to reduce sheet flexibility and thus localizes any external stress applied to the paper, so that tearing can now proceed more easily. Thirty years ago a quantitative theory of Elmendorf tear resistance was developed in terms of the work required to pull a single fiber out of the sheet during the tearing process, as opposed to the energy involved in the fiber-rupture process (*94*). Since tear resistance reflects the work required to tear a given length of paper, the tearing process should involve fiber pull-out if the structure is weakly bonded but should produce fiber rupture in a strongly bonded sheet. As a fiber becomes more frequently bonded, less stress-strain energy is required to rupture the ever-decreasing free segment length between neighboring bonded areas; hence, the work to tear the sheet passes through a maximum value and then decreases with further refining of the pulp.

The pressing of a wet fiber mat not only helps to remove some of the free water but also contributes to the ultimate strength of the dried sheet. External pressing of a web consisting of swollen, deformable fibers provides more contact area between adjacent fibers and therefore improves fiber-fiber bonding. According to Doughty, very heavy wet-pressing can greatly increase the dry tensile strength of paper (*95*). More recent work by Malmberg (*96*) tends to support Doughty's findings. Referring again to the beating curve in Figure 1.7, it appears that pressing a wet fiber mat is somewhat analogous to beating; strength properties improved by extensive beating are also generally improved by wet-pressing, while tear resistance is reduced by both of these processes. These effects result from producing a well consolidated, high density, more rigid structure having extensive bonding between fibers.

Formation of a wet fiber mat on a high speed paper machine is a process quite different from the essentially static formation of a laboratory handsheet by drainage alone. The turbulence provided by various techniques employed on commercial paper machines helps to develop a uniform fibrous mat from a relatively high consistency fiber slurry feed. A considerable amount of fiber orientation in the machine direction (MD) is produced in machine-made paper, in contrast to the completely random orientation of fibers in a laboratory handsheet (*97,98*). Some investigators have also suggested that the surface fibrils of fibers become oriented in the MD on a paper machine (*98*). In addition, the MD tension or draw required to pull the formed sheet through the paper machine is a factor not encountered in preparing laboratory handsheets (*99*). These three uni-directional machine factors combine to produce dry paper with MD tensile strength considerably higher than its cross-machine direction (CMD) strength (*97-101*). Similarly, it is easier to tear paper in the MD (i.e., with the "grain") than in the CMD (against the "grain"). The asymmetry of these strength properties in machine-made paper can be modified to some extent by adjusting paper machine operating conditions.

As discussed in an earlier section, the elevated temperature drying of a wet-pressed sheet by evaporation of most of the moisture still remaining in the paper leads to the rapid development of all dry strength properties. However, the final balance in properties (e.g., tensile vs. tear resistance) is greatly affected by the amount of restraint of the sheet during the drying process. Paper dried under considerable restraint retains built-in stresses and possesses relatively high tensile strength (in the direction of restraint), while paper dried under much less restraint develops less tensile strength but has much higher elongation to rupture (*102,103*). Page and Tydeman (*104*) have observed the production of kinks and microcompressions by the transverse shrinkage of fibers during the drying of a sheet under no restraint. These microstructure effects probably account for at least part of the reported stress-strain variations produced in paper by changing the degree of drying restraint. Perhaps an extreme qualitative example is the wrinkled, resilient paper obtained when a wet-pressed handsheet is dried under no restraint whatsoever. Paper which is encouraged to undergo a special microcreping process at 60-70% total solids is an excellent practical example of the effects produced by reducing sheet restraint during the elevated temperature drying process (*105,106*). Although this paper has less tensile strength, it shows greater elongation to rupture and exceptional stress-strain toughness and therefore has important commercial applications.

Calendering of the dried sheet is generally employed to increase surface smoothness or to densify the paper further. Wet calendering is sometimes employed to apply a surface adhesive (e.g., starch) for improving surface pick resistance or abrasion resistance. Neither wet nor dry calendering is used to increase the other strength properties of paper, and calendering may even impair these properties in some cases.

Environmental Effects

The presence of various common chemical reagents during the refining of wood pulp fibers has been reported to have certain beneficial effects on paper dry strength (*107*). However, it appears that none of these reported effects is sufficiently large, practical, or reproducible to have warranted much attention. It should be noted that the presence of certain common inorganic salts increases the measured slurry freeness and thus requires more prolonged beating if one wishes to refine the pulp to a predetermined freeness. The more heavily beaten pulp may produce paper having higher dry strength than the same pulp beaten to the same freeness in the absence of these salts, but this is not a true environmental effect.

Environmental effects during the sheetmaking process often modify the dry strength properties of paper. Accordingly, the presence of a surface active agent which lowers the surface tension of the aqueous fiber slurry reduces fiber-fiber bonding and hence may considerably reduce the dry tensile and burst strengths (*25,41,108*). Water pH, hardness, and alkalinity generally produce no significant strength effects per se. However, the presence of alum or sodium aluminate can produce aluminum hydroxide precipitate in a pulp slurry, and the adsorbed precipitate can interfere with fiber-fiber bonding and reduce paper dry strength (*68,109*). Conversely, claims have been made that certain hydroxy aluminum polymers, if prepared under the right circumstances, can greatly improve the tensile strength of paper (*110*). This latter effect does not seem to be thoroughly documented at present.

Internally Added Dry Strength Agents

The main purpose of a dry strength additive is to improve the degree of bonding between fibers in paper. Most commercial dry strength additives are hydrophilic water-soluble polymers which are added internally to the aqueous pulp slurry. The strong hydrogen bonding potential of these agents suggests the mechanism for their performance; in general, hydrophobic adhesives are ineffective dry strength additives in an aqueous papermaking system. Such materials as starches (natural and modified), various natural gums, and synthetic polymers are the primary dry strength additives currently used commercially.

The general effects of these additives have been analyzed on the same basis as the beating process (*68*), employing the modified beating curve shown in Figure 1.8 (compare with Figure 1.7). Beating which produces a shift in sheet properties (e.g., from B to C in Figure 1.8) can be considered a "mechanical hydration" effect. However, many dry strength additives can also move sheet strength properties from B to C. This change can be thought of as a "chemical hydration" effect. It is even possible to employ a dry strength additive at point A and im-

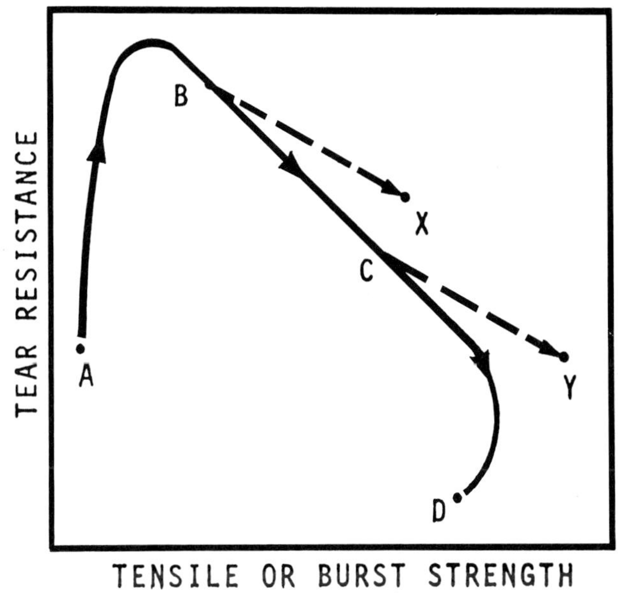

Fig. 1.8. Effects of beating and dry strength agents on strength of paper.

prove handsheet properties to those represented by point B. Thus, either pulp beating or a dry strength additive produces a tear-tensile or tear-burst curve of the ABCD type shown in Figure 1.8. Throughout the normal (linear) part of the beating curve, these effects merely improve the tensile and burst strengths of paper at the expense of its tear resistance by promoting better fiber-fiber bonding.

A few dry strength additives do more than merely take the place of extensive beating. As shown in Figure 1.8, the changes BX and CY not only improve the tensile and burst strengths but also preserve some of the tear resistance. These effects were reported some years ago for an experimental acrylic copolymer (*68*); changes BX and CY were designated as "fiber substitution" effects, because they resulted in stronger paper than could otherwise be obtained without producing a higher basis weight sheet or employing a stronger type of pulp. The modest fiber substitution effects shown in Figure 1.8 are probably due to preserving fiber length and/or strength by substituting a chemical adhesive for part of the somewhat destructive beating process.

An intensive study was conducted by Leach (*9*) to determine the effects of a self-adsorbing dry strength additive, locust bean gum, on the various parameters which govern paper dry strength. He concluded that the improvements in tensile and burst strengths produced by this natural gum should be ascribed to increased strength of the bonds (60%), improved sheet formation (25%), and increased number of bonds (15%); however, these conclusions involved indirect measurements plus deductive reasoning designed to separate the inherent bond strength and degree of bonded area parameters. As noted in an earlier section, these two factors are intimately involved as a product factor which determines the level of fiber-fiber bonding. Therefore, at this stage of our knowledge it may be prudent to consider most dry strength additives as materials which merely provide extra adhesive for fiber-fiber bonding. This may increase the degree of bonded area at each fiber-fiber crossover and/or the frequency of bonding per unit fiber length. For example, the treatment of a paper sheet with a starch solution was reported to increase both the optical contact area per crossover region and the frequency of optical contact areas along the fibers (*111*). Beating was also observed to produce both of these effects (*8*). Further studies involving dry strength additives should be informative.

Until recently, most dry strength additives have been anionic polyelectrolytes which require a cationic retention aid to be firmly adsorbed by anionic pulp fiber surfaces. Papermakers' alum has served for many years as the conventional retention aid, although cationic polymers such as polyethyleneimine (*112*) or polyamine-dibasic acid resin (*113*) have also been employed more recently. Such self-adsorbing materials as locust bean gum, cationic starches, and amphoteric polymers are also used as dry strength additives.

Effects of Other Internal Additives

Although wet strength resins are not employed primarily for their effects on the dry strength of paper, most types of commercial wet strength additives do improve dry tensile and dry burst while reducing the tear resistance of the dry sheet. Some research workers feel that certain wet strength resins produce covalent bonding between adjacent fibers in paper, thus conferring substantial wet strength to the sheet (*16,17*). It has been suggested that this covalent wet strength bonding produces an equal increase in dry strength bonding and so accounts for the observed improvement in paper dry strength (*17,18*). In addition to this concept, the addition of a wet strength agent which contributes extra hydrophilic polymeric glue should improve the dry strength of paper in the same way that a dry strength additive does. The liberal use of certain elastomeric latexes as agents for promoting wet strength in paper has been reported to improve certain dry strength properties, also (*114*).

Hydrophilic polymers are often used at low addition levels as retention or drainage aids. In general, these agents are very high molecular weight polymers which often have structures similar to some of the polymers used as dry strength additives. Accordingly, polymeric retention and drainage aids also should modestly improve paper dry strength. However, it has been suggested that the extremely high molecular weight of these agents can also encourage fiber flocculation and poor sheet formation, in which case paper dry strength may be impaired instead (*115*).

In general, predominantly hydrophobic internal additives tend to reduce paper dry strength. A prime example is internal sizing agents which are added to give paper some degree of water repellency. Commercial sizing agents are either hydrophobic materials such as wax or amphipathic (polar/nonpolar) materials such as rosin or fatty acid derivatives. Thus, the internal addition of more than 0.5% of a wax-based sizing agent significantly reduces the dry tensile and burst strengths of paper (*116*); if added at a high level, certain amphipathic sizing agents can also impair dry strength to some extent (*109*). Presumably, sizing materials adsorbed onto fiber surfaces may reduce the tendency of adjacent fibers to bond strongly during the sheet drying process. Reduced fiber-fiber bonding impairs the dry tensile and burst values, although it may improve the tear resistance of paper. If added internally at a high level, dyestuffs consisting of predominantly hydrophobic groups could have a similar effect on dry strength because of interference with interfiber bonding.

The internal addition of mineral filler particles at the high level (5-25%) generally required to obtain improved sheet opacity can severely reduce the tensile and burst strengths of paper (*109,117,118*). Although these materials are wetted by water, their particulate nature (0.1 to 2 microns) probably interferes with fiber-fiber bonding by keeping adjacent treated fiber surfaces too far apart during the sheet de-watering process. This effect should be especially severe in the many systems where filler retention involves a mechanism of particle aggregation before deposition onto fiber surfaces (*119*).

Surface Application of Additives

Dry strength additives are often applied to paper and paperboard at the size press or calender stack in order to improve surface strength properties such as abrasion resistance and pick resistance. A primary effect of these agents is to bond protruding surface fibers down onto the underlying fibrous mat. In general, hydrophilic polymers, such as starches, carboxymethyl cellulose, and polyvinyl alcohol, are employed (*120*). To the extent that these surface strengthening agents penetrate to the interior of the sheet, the strength properties (especially internal bond strength) may also be improved; the added adhesive can lead to increased bonding between the already bonded fibers. Sheet stiffness or rigidity may also be improved by the surface application of dry strength agents. Similar hydrophilic polymers are often applied to paper surfaces to "size" the sheet against nonpolar fluids (e.g., molten wax, printing inks) to be applied later. Here the main purpose is to prevent deep penetration of hydrophobic liquids, but surface strength properties may also be improved.

In principle, the external addition (surface application) of hydrophobic materials to paper should not interfere with the interfiber bonds already formed during the papermaking process. However, severe rewetting of the dried sheet during the surface application of an aqueous dispersion of a hydrophobic material may impair the ultimate dry strength of paper. Therefore, some care must be exercised to avoid this adverse effect during the surface application of water-based sizing agents, latexes, and coating colors.

Effects of Saturants

The addition of a very large amount of material to heavily impregnate or saturate the sheet is usually of interest only in the preparation of specialty grades of paper and paperboard, where properties other than dry strength are of prime importance. Impregnating agents include waxes, polymeric latexes, and flame retardants. Saturants may be added internally to a pulp fiber slurry; thus, the internal addition of 5-10% elastomeric latex for wet strength purposes also improves certain dry strength properties (*114*). Generally, however, impregnating agents are added to the formed sheet to obtain readily the desired high addition level of 10-50%, based on fiber weight. When the finished paper product contains such a large amount of additive material, the sheet's dry strength properties begin to lose their original significance. There is the question of whether the strength of the product is due primarily to its bonded cellulose fiber skeleton or to its massive amount of non-cellulosic component. In the extreme, a heavily saturated paper becomes a fiber-reinforced resin or plastic structure.

Literature Cited

1. C.T.J. Dodson, *Rep. Prog. Phys.*, 33(1): 1 (1970).
2. R. E. Benson, *Tappi,* 54(5): 699 (1971).
3. B. Radvan, C. Dodson, and C. G. Skold, In "Consolidation of the Paper Web," ed. by F. Bolam, Technical Section, BPBMA, London, 1966, pp. 189-214.
4. G. Haywood, *Tappi*, 33(8): 370 (1950).
5. G. Jayme and G. Hunger, In "Formation and Structure of Paper," ed. by F. Bolam, Technical Section, BPBMA, London, 1962, pp. 135-170.
6. G. L. Forgacs and D. Atack, In "Formation and Structure of Paper," ed. by F. Bolam, Technical Section, BPBMA, London, 1962, pp. 721-737.
7. R. W. Davison, *Tappi*, 54(4): 567 (1972).
8. D. H. Page, P. A. Tydeman, and M. Hunt, In "Formation and Structure of Paper," ed. by F. Bolam, Technical Section, BPBMA, London, 1962, pp. 171-193.
9. H. J. Leech, *Tappi*, 37(8): 343 (1954).
10. R. J. Norman, In "Consolidation of the Paper Web," ed. by F. Bolam, Technical Section, BPBMA, London, 1966, pp. 269-298.
11. O. Andersson and I. Steen, In "Formation and Structure of Paper," ed. by F. Bolam, Technical Section, BPBMA, London, 1962, pp. 771-782.
12. R. H. Doughty, *Pap. Trade J.*, 93(2): 15 (1931).
13. J. d'A. Clark, *Tech. Assoc. Pap.*, 26: 499 (1943).

14. J. Strachan, *Proc. Tech. Sect. BPBMA*, 6(2): 139 (1926).
15. W. S. Wilson, *Tappi*, 34(12): 561 (1951).
16. C. G. Landes, *Tappi*, 33(9): 463-71 (1950).
17. D. J. Salley and A. F. Blockman, *Tech. Assoc. Pap.*, 30: 223 (1947).
18. A. Jurecic, T. Lindh, S. E. Church, and V. Stannett, *Tappi*, 41(9): 465 (1958).
19. G. G. Allan and W. M. Reif, *sven. papperstidn.*, 74(18): 563 (1971).
20. W. M. Reif, "The Utilization of Ionic Bonds in Development of Paper Strength," paper presented at TAPPI Papermakers Conf., Atlanta, June 1972, pp. 47-61.
21. H. Mark, In "Cellulose and Cellulose Derivatives," ed. by E. Ott, Interscience, New York, 1943, pp. 999-1004.
22. G. L. Clark, *Ind. Eng. Chem.*, 22: 474 (1930).
23. J. d'A. Clark, *Pap. Trade J.*, 97(26): 25 (1933).
24. L. G. Cottrall, *Tappi*, 33(9): 471 (1950).
25. G. Broughton and J. P. Wang, *Tappi*, 38(7): 412 (1955).
26. J. A. Van den Akker, *Tappi*, 35(1): 13 (1952).
27. R. H. Marchessault, W. C. Lodge, and S. G. Mason, *sven. papperstidn.*, 59(24): 859 (1956).
28. H. Corte and H. Schaschek, *Das Papier*, 9: 519 (1955).
29. D. H. Page, *Pap. Technol., 1(4): T165 (1960).*
30. D. H. Page and P. A. Tydeman, *Pap. Technol.*, 1(5): T207 (1960).
31. D. H. Page, P. A. Tydeman, and M. Hunt, In "Formation and Structure of Paper," ed. by F. Bolam, Technical Section, BPBMA, London, 1962, pp. 249-263.
32. J. E. Stone, *Pulp Pap. Mag. Can.*, 64(12): T528 (1963).
33. O. J. Kallmes and C. Eckert, *Tappi*, 47(9): 540 (1964).
34. A. H. Nissan and S. S. Sternstein, *Tappi*, 47(1): 1 (1964).
35. C. H. Mayhood, Jr., O. J. Kallmes, and M. M. Cauley, *Tappi*, 45(1): 69 (1962).
36. A. P. Schniewind, L. J. Nemeth, and D. L. Brink, *Tappi*, 47(4): 244 (1964).
37. D. C. McIntosh, *Tappi*, 46(5): 273 (1963).
38. U-B. Mohlin, *sven. papperstidn.*, 77(4): 131 (1974).
39. W. B. Campbell, *Pap. Trade J.*, 95(8): 29 (1932).
40. B. Steenberg, N. Thalen, and D. Wahren, In "Consolidation of the Paper Web," ed. by F. Bolam, Technical Section, BPBMA, London, 1966, pp. 177-186.
41. L. M. Lyne and W. Gallay, *Pulp Pap. Mag. Can.*, 55: 128 (Oct. 1954).
42. A. A. Robertson, In "Consolidation of the Paper Web," ed. by F. Bolam, Technical Section, BPBMA, London, 1966, pp. 90-115.
43. D. H. Page and P. A. Tydeman, In "Consolidation of the Paper Web," ed. by F. Bolam, Technical Section, BPBMA, London, 1966, pp. 371-392.
44. J. E. Stone and A. M. Scallan, In "Consolidation of the Paper Web," ed. by F. Bolam, Technical Section, BPBMA, London, 1966, pp. 145-166.
45. J. A. Van den Akker, A. L. Lathrop, M. H. Voelker, and L. R. Dearth, *Tappi*, 41(8): 416 (1958).
46. B. A. Jayne, *For. Prod. J.*, 10(6): 316 (1960).
47. D. C. McIntosh and B. Leopold, In "Formation and Structure of Paper," ed. by F. Bolam, Technical Section, BPBMA, London, 1962, pp. 265-270.
48. B. Leopold and D. C. McIntosh, *Tappi*, 44(3): 235 (1961).
49. N. Hartler, G. Kull, and L. Stockman, *sven. papperstidn.*, 66(8): 301 (1963).
50. J. G. Buchanan and O. V. Washburn, *Pulp Pap. Mag. Can.*, 63(10): T485 (1962).
51. D. H. Page, *Tappi*, 52(4): 674 (1969).
52. H. Corte, O. J. Kallmes, and D. Jarrot, *Paper Maker,* 142(7): 61 (1961).
53. T. Helle, *sven. papperstidn.*, 66(24): 1015 (1963).
54. D. Graham, *Tappi*, 39(3): 147 (1956).
55. J. d'A. Clark, *Tappi*, 56(7): 122 (1973).
56. A. R. Jones, *Tappi*, 55(10): 1522 (1972).
57. J. d'A. Clark, *Tech. Assoc. Pap.*, 25: 556 (1942).
58. "Formation and Structure of Paper," ed. by F. Bolam, Technical Section, BPBMA, London, 1962, discussion pp. 565-572.
59. V. Balodis, *Aust. J. Appl. Sci.*, 14(4): 284 (1963).

60. O. Andersson and O. Falk, *sven. papperstidn.*, 69(4): 91 (1966).
61. J. Goldschmidt and D. Wahren, *sven. papperstidn.*, 71(13-14): 477 (1968).
62. R. S. Seth and D. H. Page, *J. Mater. Sci.*, 9(11): 1745 (1974).
63. A. H. Nissan, *Tex. Res. J.*, 25: 780 (1955).
64. A. H. Nissan, *Tappi*, 39(2): 93 (1956).
65. S. S. Sternstein and A. H. Nissan, In "Formation and Structure of Paper," ed. by F. Bolam, Technical Section, BPBMA, London, 1962, pp. 319-49.
66. J. A. Van den Akker, *Tappi*, 53(3): 388 (1970).
67. W. H. Algar, In "Consolidation of the Paper Web," ed. by F. Bolam, Technical Section, BPBMA, London, 1966, pp. 814-49.
68. R. W. Davison, S. T. Putnam, R. T. Mashburn, and H. O. Ware, *Tappi*, 40(7): 499 (1957).
69. J. Vilars, *Papeterie*, 71: 127 (1949).
70. S. M. Temple, *Proc. Australian Pulp Paper Ind. Tech Assoc.*, 2: 132 (1948).
71. G. Haywood, *Tappi*, 33(8): 370 (1950).
72. H. F. Rance, *Paper Maker*, Midsummer issue :23 (1949).
73. D. H. Page, F. El-Hosseiny, K. Winkler, and R. Bain, *Pulp Pap. Mag. Can.*, 73(8): 72 (1972).
74. H. W. Giertz, In "Formation and Structure of Paper," ed. by F. Bolam, Technical Section, BPBMA, London, 1962, pp. 597-620.
75. R. C. McKee, *Pap. Trade J.*, 155(21): 34 (1971).
76. H. G. Higgins, A. W. McKenzie, and K. J. Harrington, *Tappi*, 41(5): 193 (1958).
77. J. C. Bletzinger, *Ind. Eng. Chem.*, 35: 474 (1943).
78. W. H. Aiken, *Ind. Eng. Chem.*, 35: 1206 (1943).
79. J. J. Harrison, *Pap. Trade J.*, 119(5): 28 (1944).
80. J. A. Walecka, *Tappi*, 39(7): 458 (1956).
81. K. K. Talwar, *Tappi*, 41(5): 207 (1958).
82. J. A. Harpham, A. R. Reid, and H. W. Turner, *Tappi*, 41(11): 629 (1958).
83. J. J. Spadaro, H. J. Janssen, H.L.E. Vix, and F. L. Simons, *Tappi*, 41(11): 674 (1958).
84. R. H. Gillespie, M. Mueller, W. Swenson, and K. Ward Jr., *Tappi*, 44(9): 662 (1961).
85. D.A.I. Goring, *Pulp Pap. Mag. Can.*, 68(8): T372 (1967).
86. P. F. Brown and J. W. Swanson, *Tappi*, 54(12): 2012 (1971).
87. D.A.I. Goring and G. Suranyi, *Pulp Pap. Mag. Can.*, 70(20): T390 (1969).
88. G. Centola and D. Borruso, *Tappi*, 50(7): 344 (1967).
89. F. Mobarak, A. E. El-Ashmawy, and Y. Fahmy, *Cellul. Chem. Technol.*, 7: 325 (1973).
90. D. W. Clayton and J. E. Stone, *Pulp Pap. Mag. Can.*, 64(11): T459 (1963).
91. E. Strazdins, *Tappi*, 55(12): 1691 (1972).
92. H. C. Higgins and J. De Yong, In "Formation and Structure of Paper," ed. by F. Bolam, Technical Section, BPBMA, London, 1962, pp. 651-90.
93. O. Kress and H. Bialkowsky, *Pap. Trade J.*, 93(20): 35 (1931).
94. Staff of Inst. Paper Chem., *Pap. Trade J.*, 118(5): 13 (1944).
95. R. H. Doughty, *Pap. Trade J.*, 95(10): 111 (1932).
96. B. Malmberg, *sven. papperstidn.*, 73(10): 334 (1970).
97. A. M. Lund, *Pap. Trade J.*, 99(6): 37 (1934).
98. W. Gallay, In "Formation and Structure of Paper," ed. by F. Bolam, Technical Section, BPBMA, London, 1962, pp. 491-532.
99. B. Steenberg, *Pulp Pap. Mag. Can.*, 50(3): 207 (1949).
100. W. A. Selke (to Kimberly-Clark), U.S. Pat. 3,205,124 (Sept. 7, 1965).
101. J. E. Sapp and W. F. Gillespie, *Pap. Trade J.*, 124(9): 120 (1947).
102. S.R.H. Edge, *Proc. Tech. Sect. BPBMA*, 26: 425 (1945).
103. J. H. Schulz, *Tappi*, 44(10): 736 (1961).
104. D. H. Page and P. A. Tydeman, In "Formation and Structure of Paper," ed. by F. Bolam, Technical Section, BPBMA, London, 1962, pp. 397-413.
105. H. S. Welsh, In "Consolidation of the Paper Web," ed. by F. Bolam, Technical Section, BPBMA, London, 1966, pp. 397-409.
106. E. M. Burrow, *Pap. Technol.*, 6: T125 (1965).
107. J. P. Casey, "Pulp and Paper Chemistry and Chemical Technology," 2d ed., Interscience, New York, 1960, vol. 2, pp. 635-38.

108. R. V. Touchette and L. C. Jenness, *Tappi*, 43(5): 484 (1960).
109. H. C. Schwalbe, In "Consolidation of the Paper Web," ed. by F. Bolam, Technical Section, BPBMA, London, 1966, pp. 692-734.
110. P. Lagally and H. Lagally, *Tappi*, 42(11): 888 (1959).
111. D. H. Page, In "Consolidation of the Paper Web," ed. by F. Bolam, Technical Section, BPBMA, London, 1966, discussion pp. 737-38.
112. S. Machida, T. Yoshii, and H. Yamada, *JAPAN TAPPI*, 19(7): 353; (8): 389 (1965).
113. W. F. Linke, *Tappi*, 51(11): 59A (1968).
114. D. Graham, *Tappi*, 41(5): 231 (1958).
115. W. F. Reynolds, *Tappi*, 57(3): 116 (1974).
116. R. W. Davison and H. M. Spurlin, In "Handbook of Pulp and Paper Technology," ed. by K. W. Britt, 2d ed., Van Nostrand Reinhold, New York, 1970, pp. 355-368.
117. D. A. Hughes, *Tappi*, 45(2): 159A (1962).
118. W. V. Arvold, Jr. and P. Calott, *Tappi*, 39(11): 823 (1956).
119. K. W. Britt, *Tappi*, 56(10): 46 (1973).
120. E. J. Barber and H. O. Ware, *Tappi*, 40(5): 365 (1957).

2

Surface Applications

Chester T. Beals

INTRODUCTION

Swanson (*1*) presents an interesting insight into the importance to the paper industry of chemical additives. "Instead of the multitude of grades of paper and board presently consumed in the amount of more than 54 million tons, production would be confined to un-colored, non wet-strength tissues, a few absorbant papers, some glassine, unsized writing paper and perhaps unsized newsprint. Total production might approximate 5 million tons. Thus by incorporating about 5 million tons of various chemicals, the paper industry has attained a market ten times as large" (*1*).

In this chapter the uses of additives applied to the surface of the sheet are examined. The materials can be put on at various points on the paper machine, although by far the most common method is application to the sheet after almost complete drying. Very often surface additives are used for purposes other than direct strength improvement. However, strength improvement is inherent in almost every type of surface treatment and usually accrues as an additional benefit which may be of considerable economic value.

Natural starches are easily the most prevalent surface additives, and the machine size press is by far the most typical point of addition. It is beyond the scope of this chapter to deal in detail with the chemistry of starches or with the design and operation of the equipment for application. Rather, this chapter concerns itself only with such background information and pertinent data as are of interest to the person working directly with these processes. For the reader who wishes further information, references are to be found in the volumes prepared by Casey (*2*), Kerr (*3*), Whistler and Paschall (*4,5*), Macdonald and Franklin (*6*), and the Corn Industries Research Foundation (*7*). TAPPI Monograph No. 17 (*8*) includes a summary of starch related data that is of much value in relationship to surface additives, as well as to coatings.

PURPOSES

The commonly used term "surface sizing" applies to the process as old as papermaking itself of applying such adhesive and filming agents as animal glue, starch pastes, and CMC to the essentially dried sheet of paper. The term "sizing" implies resistance to water and, as such, is something of a misnomer as a reference to a surface treatment, at least with most conventional additives. The degree of water penetration as measured by conventional size tests is usually not affected to any great degree. However, surface treatments do indeed develop a surface resistance to the absorption and spread (feathering) of conventional writing inks, as well as to the penetration of oil base printing inks. Additionally, in the manufacture of sheets for offset printing, the combination of a surface treatment with a suitable degree of internal sizing prevents softening of the surface when moistened on the press (9).

Surface treatments carry several important advantages over the furnish additives that may be used for similar general effects. First, a generally higher level of treatment can be gained without the limitations of retention that are inherent in furnish addition. In fact, the only conditions detracting from 100% utilization are such losses as spillage and cleanups. By this token, much greater gains not only in strength values but also in enhanced surface characteristics can be expected. Second, surface treatments afford a greater control as to location in the sheet, i.e., surface or internal absorption. Third, surface treatments often offer acceptable alternatives to wet end treatments that may aggravate such problems as reduced wire drainage, plugging of wires or felts, excessive slime buildup, or poor retention. Finally, surface treatments provide a final control of sheet quality that may be fairly independent of the wet end operations of the paper machine, although it must be kept in mind that a base sheet of proper quality is extremely important to the effectiveness of the treatment.

The main disadvantage of a surface treatment, particularly at the size press, is the limitation in machine productive capacity because of the need to dry the moisture that is added with the size solution. The costs of such reduced production or of the additional drying capacity required to maintain production must, of course, be justified by the advantages gained.

Specific improvements gained by surface treatments are in structural strength qualities as measured by mullen, tensile, Scott internal bond, Z-direction tensile and fold; in surface strength qualities as measured by IGT, Dennison wax pick, various linting tests, scuff resistance, and erasability; and in many additional sheet qualities that are relatively non-related to strength. These include reduced porosity; increased stiffness; improved surface smoothness and printability, particularly with gloss inks; and resistance to grease, oil, or varnish. Base sheet qualities for coating stock can be enhanced, and many strictly functional qualities such as surface release can be imparted to specialty grades.

Within the limitations of this chapter, the first two categories are of primary interest. Many surface applications are made specifically for the purposes outlined in the third category. However, strength improvements almost inevitably accrue to important economic advantage, even though sheet strength is not of primary concern. It may be that less expensive fiber or increased filler can be used in the furnish, or refining can be reduced to allow improved drainage and formation, easier drying of the base sheet, and a final sheet with better dimensional stability.

SURFACE SIZING MATERIALS

The two prime requirements for a satisfactory surface additive are good adhesive qualities and fair to excellent film forming characteristics. The relative balance of these qualities, of course, largely determines the product to be chosen for a particular need. Cost is important, although very costly materials may be used if the superior qualities gained justify the expense. Other considerations include compatibility with the general process, ease and safety of handling, stability during storage and application, and, most important, compliance with FDA regulations for the intended use.

The products that are available to the industry can be grouped into three general categories.

Of major importance as strengthening agents are starch and animal glue. Starch is the most widely used product for this purpose because of its low cost, ready availability, and good performance in the great majority of paper grades. Starch in its native form is a relatively poor film former. In fact, it is reported that a truly continuous film is rarely attained on a sheet of paper (*9*). However, starch derivatives are available that rival the more expensive products noted below that are used specifically for this property. The bulk of this chapter is devoted to starch as the typical surface strength additive because of its very wide use. Animal glue is an excellent film former and strengthening agent. However, its high cost limits its use to very high grade fine papers where surface strength and resistance to erasure are of prime importance. This product imparts a relatively tough non-absorbent surface that can be made water resistant by passing it through a bath of alum or formaldehyde. Glue also improves strength, particularly the folding endurance of the sheet (*10*).

In the second major category, superior film formers, used mainly for such qualities as oil or grease holdout or improved printing qualities, include CMC, methyl cellulose, polyvinyl alcohol, styrene-malaic anhydride copolymers, and alginate. These products are relatively expensive. They can be used alone but more often are incorporated with starch or other materials.

The final category includes wax emulsions, synthetic sizing agents, and truly exotic products such as stearato chromic chloride and fluorochemicals. These are employed to impart a high degree of water resistance or oil resistance in the case of the latter to the sheet.

Other adjuncts are often used in surface size formulations as required by local conditions. These may be necessary to give special properties to the films or to control inherent operating problems. None has a major direct bearing on strength development efficiency. These products include defoamers; slimicides; plasticizers; anti-stick agents such as soaps, glycols, or wax emulsions; and anti-static agents. Pigments such as clay or clay-titanium may be incorporated to form a low solids wash coating; however, this is done solely to improve optical or precoat properties and again is not related to strength.

STARCH AS A STRENGTH ADDITIVE

The importance of starch in the paper industry can be appreciated only when it is realized that the paper and paper processing industries are the largest industrial users of starch and its derivatives. Alternately, starch ranks second only to clay in the quantity consumption of non-fibrous raw materials for papermaking. Starch performs its function as an adhesive; that is, it creates a bond between units in a paper structure. The bond may be within the fiber structure of the sheet itself, for pigment in a coating, or between the components of a container. Approximately 50% of the starch in the industry is used as a paper surface size, the remainder being split about equally between coating on the one hand and corrugating, laminating, or general adhesive uses on the other. Corn starch comprises about three quarters of the total available from all specie sources (*1*).

Starch is manufactured commercially from several sources. The four most important specie types used in the United States are corn, potato, tapioca, and wheat, with corn as the typical raw material source.

Potato and tapioca starches represent the tuber or root type products, which differ in some basic properties from the cereal types. Potato starch is produced domestically and is also available as an import from the Netherlands. Tapioca is produced exclusively in the tropics, and at present the chief supply points are Thailand and Brazil. Both availability and prices of both of these products are subject to much wider fluctuations than are those of corn, and for this reason relative usages generally follow existing market conditions. At times in the past, both products have been able to compete very well with corn starch, particularly at locations near ports of entry or primary production areas. Recent advances in the manufacture of prepared potato products in the food industry have greatly reduced the supply of "cull grade" potatoes that were formerly unacceptable for food purposes. This has increased costs to the starch processor and has in turn limited the supply of potato starch for industrial use. However, some derivatized potato products do continue to maintain a position because of certain superior qualities over corn, as noted below.

There are several basic differences between potato starch and corn. Potato starch has a larger granule size of somewhat higher density, which results in different handling characteristics as a slurry. This offers few problems as long as proper handling methods are used. The gel temperature of potato starch is significantly lower; this may be of advantage to some users, depending upon their particular requirements for cookout. Kofler gel temperature ranges are listed (*11*) as 56° - 66°C (133° - 151°F) for potato, 58.5° - 70°C (137° - 158°F) for tapioca, and 62° - 72°C (144° - 162°F) for corn starch. Molecular chain length is longer by double or more (*12*) and gives stringier, more cohesive viscosity characteristics to the cooked out pastes. Derivatized potato starches that are superior to similar corn base products in the formation of high solids surface films are prepared, and these products have been able to compete with other non-starch film formers in the manufacture of high quality functional coatings.

Wheat starch is available in the United States in limited quantities, as dictated by a refining industry geared to the production of other wheat products, mainly gluten. Wheat starch, like potato starch, has a low gel temperature, 52° - 63°C (126° - 145°F) (*11*), and also has excellent film forming characteristics. This well refined product has been found to be interchangeable with corn starch for most paper industry applications.

A starch product is differentiated from a flour in that starch is refined for removal of all but a minimal trace of protein and various other components of the parent product. For example, a normal corn starch contains 0.4% or less protein, as compared with 9% in the corn kernel (7). (A discussion of the importance of efficient refining to viscosity stability follows in a later chapter.) This requirement is as important for wheat starch as it is for corn.

It is interesting to note work with wheat flour by research groups seeking new uses in the industry for wheat products. Rankin et al. (*13*) compared acid modified wheat flour containing 9% protein (dry solids) with a commercial hypochlorite oxidized starch on both bond and unbleached bag paper grades on the size press of a small commercial fourdrinier machine. After one 5½ hour trial run, they found that strength and surface properties were comparable and that there were no operating troubles, although there was some buildup in protein which seemed to be approaching equilibrium near the end of the trial. In another work, Rankin et al. (*14*) reported the results of lab scale tests in which, "Surface sizes for paper were prepared by enzyme conversion of cationic aminoethyl flours and starches. Aminoethylation of corn and wheat flour and starch was accomplished with ethylenimine. In laboratory surface sizing studies, the sizes were comparable to commercial cationic starch sizes in improving paper strength and surface properties. All of the cationic products were highly retained when the paper was recycled" (*14*). Austin and Sullivan (*15*) reported successful work with wheat flour containing as high as 5% protein used as a size press treatment on commercially run 42 pound linerboard. They found that considerable savings could be realized. However, for flours above 0.5% protein, a double conversion with both starch and protein liquefying enzymes is necessary. The two conversion operations require different conditions and cannot be carried out concurrently.

Granular Characteristics

Starch manufactured for sale exists in the form of granules that have been separated, unaltered in form, from the parent cereal grain or tuber. The product is white or pale buff colored and may be in either a powder or the "pearl" form of soft agglomerates of granules produced in the drying process. Microscopic examination reveals several physical properties that are unique to each specie source. Most readily apparent are the characteristic granule sizes and shapes. When viewed under polarized light, the position of the hilum, presence or absence of striations, and the property of birefringence all become apparent. The hilum is the point in the granule which is the organic center, or nucleus, around which the granule is grown. Some starches show a series of markings arranged concentrically around the hilum. These are known as the striations and are most apparent in the potato starch granule. Birefringence is the most striking characteristic that becomes apparent under polarized light. This appears as a dark cross within the otherwise illuminated granule, the intersection being located at the hilum. The shapes and positions of all of these markings vary with specie type and afford a ready means of identification for at least the common species of concern in this discussion (*16*).

Starches show a wide variation in granule sizes both within a specie and between species. Potato starch granules are the largest, ranging in size from 15 to 100 microns and averaging 33 microns. Tapioca and wheat rank next, each averaging 20 microns. Corn starch granules range from 5 to 25 microns and average 15. Rice starch granules are the smallest and most consistent in size, ranging from 3 to 8 microns (*17*). These dimension variations assume importance when one considers equipment for handling the dry or slurried products.

Paste Characteristics

The D-glucose molecule forms the basic building block for the starch molecule, and in this respect all starches are alike, regardless of source. However, the glucose units are arranged into two forms of polysaccharide molecules, the straight chain or amylose fraction and the branched chain or "bush" form known as amylopectin. Both the lengths of the molecules (degree of polymerization) and the relative amounts of amylose to amylopectin vary among plant sources and so impart varying properties to the starch pastes. Again comparing corn with potato starch, Kerr (*12*) lists the average molecular lengths for the amylose and amylopectin fractions of corn starch as DP 480 and 1450 (number of glucose units), respectively, while those for potato starch are 850 and greater than 2000. Amylose forms approximately 29% and 25% of the polysaccharides in corn and potato starch, respectively. Wider differences occur in other starch types not important to the paper industry. One further difference between fractions that is of interest is the staining characteristics with iodine. Amylose stains a bright blue, and amylopectin stains a purple or reddish color (*18*). TAPPI Monograph Series No. 3 (1947) indicates the following differences between amylose and amylopectin.

Table 2.1
Comparative Differences between Amylose and Amylopectin

	Amylose	Amylopectin
Shape of Molecular	Linear	Branched
Molecular Weight	10,000-60,000	50,000-1,000,000
Tendency to Retrograde	Strong, Rapid	Weak or Lacking
Action of Beta-Amylase Enzyme	Almost 100% Hydrolyzed	Only 60% Hydrolyzed
Behavior Towards H_2O	Solid and Liquid Phases Separate	Forms Paste
Color With Iodine	Deep Blue	Reddish Brown

The generally accepted hypothesis for the arrangement of the molecules within the granule shows a development of concentric layers, with the molecules in each layer being closely packed and oriented in the radial direction. The straight and branch chains are bound together at many points by primary valences forming bundles or "micelles" (*18,19,20*). In the natural state, starch is insoluble in cold water but can absorb up to 25-30% of its weight without swelling. As a slurry is heated, water is able to penetrate the granule, dissociate the hydrogen bonds that hold the micellar structural units together, and begin to hydrate the many hydroxyl groups along the length of the molecular chains.

The granule breaks down in three distinct steps. In the first phase, water is absorbed with only limited swelling of the granule. The second phase starts within a narrow range of temperature (gel temperature range) which is characteristic of a species. The granule suddenly swells, increasing many times in size, taking up a large proportion of water, and rapidly losing its birefringence. Viscosity increases rapidly during the second phase of swelling. As the temperature rise continues, the cookout enters the third phase, during which the granules begin to disintegrate. Soluble starch molecules are leached out and dispersed within the water medium, and viscosity decreases at a rate hastened by mechanical action. The large amounts of water that can be associated with starch during cookout have been reported by Kerr as 300%, about 1000%, and up to 2500%, based on original weight, at 60°C (140°F), 70°C (158°F), and maximum swelling, respectively (*19,21*).

The disappearance of the birefringence of the granule has been found to be the most precise means of pinpointing the gel temperature. The Kofler temperature ranges quoted earlier in this chapter were determined by observing the disappearance of the crosses in individual granules when heated on the hot stage of the Kofler microscope. Gel temperature is reported as a range, rather than a specific reading, since granules of different sizes begin to swell at slightly different temperatures (*11*).

The cooked out paste, in which the starch molecules are completely dispersed from the granules, assumes a viscosity that is dependent upon several conditions, most important of which are molecular length, molecular dispersion, paste concentration, and temperature. Molecular length, of course, depends upon the specie source in the case of untreated (native) starches and on the type and degree of molecular de-polymerization in the chemically treated (modified) or enzyme or chemically converted grades. Molecular dispersion generally depends upon the degree of agitation to which the paste is being or has been subjected. Many of the micellar bundles do not break up during cookout under mild agitation. Final viscosity of such a paste is much higher than that of a paste subjected to, for example, the violent agitation of an electric blender. Paste concentration affects viscosity as a function of molecular dispersion, and temperature has an inverse effect on viscosity due to the energy imparted to the system.

When a cooked out starch paste is allowed to age, molecular reassociation (set back) and retrogradation (recrystallization) progress (*21,22*) at rates which are generally dependent upon the conditions noted above. Viscosity increases to the extremes of a hard gel at high concentrations and moderate temperatures. Under certain conditions retrogradation progresses to the formation of a microcrystalline precipitate similar in appearance to a native starch granule, even to the extent of showing birefringence. This phenomenon is the result of hydrogen bonding between starch molecules that have both hydroxyl groups and hydrogen acceptor sites. The extended linear molecules of amylose are much more prone to retrogradation than are those of highly branched amylopectin, since they are freer to orient themselves with respect to other amylose molecules in solution. The great number of available hydrogen bonding sites results in gel formation and, under favorable conditions, in the formation of numerous crystallites which aggregate to form the granular precipitate. In fact, one of the means of separating the amylose and amylopectin fractions of starch is the deliberate retrogradation and precipitation of the former. (*23*)

Starch Modification and Derivatization

Discussion of the various physical and chemical treatments and handling techniques are confined to corn starch, since this product makes up 75% or more of the total usage of all varieties. In reality, only relatively minor quantities of starches other than corn are sold in other than the native or "unmodified" form. In this section the two general areas in which the manufacturer chemically treats the refined starch are considered. These are termed "modification" where the molecules are shortened for viscosity control and "derivatization" where chemical groups are attached to the molecules to impart special properties. A third area of treatment, "crosslinking," provides cross bonds between molecular groups. This treatment increases the viscosity of the cooked pastes and is not generally applicable to products for surface application for dry strength development.

Cooked out dispersions of native starches have the highest inherent molecular strength and so can be expected to be the most efficient ones for imparting dry strength to the sheet. However, such dispersions are highly viscous even at low solids levels and are prone to excessive set back on aging or even moderate cooling. In practice, an unmodified corn starch paste of over three or four percent concentration is difficult to manage for either size press or calender use. Since the amount of material that can be applied to the sheet depends largely on the solids concentration, such dispersions cannot be expected to give appreciable internal strength benefits, although they can be used as a light calender treatment for surface effects. In this capacity they lay surface fibers and improve surface strength to a limited extent. However, such uses are confined to heavier weight sheets and boards. It follows then that the starch must be modified to reduce viscosity and so provide free-flowing cooked out pastes at higher concentrations. All treatments for modification depend upon the controlled chemical depolymerization of the long chain molecules to shorter units. Such depolymerization reduces the effectiveness of the starch for strength development, so that selection of a product for a given application depends upon the balance between the viscosity limitations of the process and the strength and surface requirements for the starch.

Of the several methods available for viscosity reduction, three processes are regularly used by the manufacturer prior to sale of the product. These are acid modification, oxidation, and "torrefication," or dry roasting. The latter process is worthy of mention only because the end products are dextrins which are used only rarely, if at all, as surface additives on the paper machine.

Acid Modified Starches. Acid modification is the most economical means for viscosity reduction available to the manufacturer. Native starch in the slurry form is treated with either hydrochloric or sulfuric acid and is allowed to react at a temperature below the gel point of the starch. The acid concentration and reaction time determine the degree of molecular breakdown. Since the starch granules remain unswelled, the neutralized, washed, and dried end product is identical in physical appearance to the native starch. These products are known in the trade as acid modified or "thin boils." They exhibit excellent viscosity stability against the shear forces to which they are subjected during their application on the size press; however, the character of the acid reaction is such that molecules are cleaved into a relatively large number of straight chain fragments that are prone to reassociation and retrogradation (*24*). Film forming characteristics are relatively poor. Although low in cost, "thin boils" are used only in paper applications in which their tendency to excessive setback does not give problems. The competition from the in plant prepared products covered in a later section has greatly limited the quantities of acid modified starches purchased for these purposes. Their main popularity is for food, textile, and adhesive uses.

Oxidized Starches. Oxidized starches are somewhat more costly to manufacture but are much more acceptable to the paper manufacturer as a surface sizing agent or coating binder. Oxidation is normally carried out by reaction with sodium hypochlorite, again at a temperature below the gel point. Products of this type are also known as "chlorinated" starches; however, the term is a misnomer since no chlorine is actually reacted to the starch molecule. Oxidized starches, in contrast to "thin boils," produce pastes of better film forming qualities, excellent stability against setback, and high clarity. They are not used more widely for two reasons — their relatively high cost and, possibly a more serious drawback, their property of being highly efficient dispersing agents toward pigments and fillers. The impact of the latter in the paper industry is discussed more fully later. Oxidized starches are used to a rather limited extent in both size press applications and coatings by mills which are willing to trade these drawbacks for the advantages offered by an excellent pre-modified uniform product.

Three additional processes for viscosity reduction are available to the customer and are discussed in the section dealing with handling and preparation in the mill. These are high energy cooking, enzyme converting, and chemical converting.

Derivatized Starches. It is stated that starch chemistry did not come of age until about 1940 (*25*). Since that time many processes have been developed for attaching chemical groups or otherwise altering starch molecules to impart special properties for specific end uses. The term "starch derivatives" is applied to this overall class of products. They should not be confused with the modified starches, since in most processes the molecular chains are not degraded to any great extent. If viscosity reduction is necessary, the products must be so treated by a separate step.

Some of the most common starch derivatives available to the papermaker as surface additives for dry strength improvement include hydroxyethyl, acetate, nitrile, and cationic starches. All are developed for enhanced properties, some examples of which are easier handling, reduced setback tendencies, improved water holding properties, and stronger and more continuous films. All treatments involve additional costs which must be justified by the economical advantages to the user. Large tonnages are in use in the paper industry, chiefly on grades requiring a high surface strength as exemplified by fine papers and high quality offsets. All of these products are reacted in the slurry form at temperatures below the gel point, so that the granules remain in the same physical form as the native starch.

The hydroxyethyl derivatives, also known as ethylated starches, are the ones most widely used in the paper industry for size press and calender treatments, as well as in specialized coatings. These products are manufactured by reacting slurried starch in alkaline medium at not over 50°C (approximately 120°F) with ethylene oxide. The level of substitution is low, being controlled to attach a hydroxyethyl group to a range of every 10th to 20th D-glucose unit. Many of the properties of the parent starch are altered to an increasing extent, depending upon the level of substitution. Granule swelling (gel) temperature is lowered. In fact, at substitutions materially above the range noted above, the granules partially swell even at the reduced reacting temperature. Swelling can be inhibited by the presence of a neutral alkali metal salt (normally NaCl). Finished product gel temperature may be as low as 60°C (140°F). The hydroxyethyl groups attached to the molecule very effectively inhibit the tendency to reassociate or retrograde and result in excellent viscosity stability even at high concentrations. The granules on cookout absorb and carry a greater amount of water, which results in continuous films of high clarity and toughness. The ether linkages are extremely stable, so that the products can be modified to high fluidities without reducing the substituent groups (*26*). From a further chemical standpoint, the ethylated derivatives are superior in reactivity to such other chemicals as cross linking or insolubilizing agents.

Ethylated starches are available at several substitution levels. The advantages noted above, as well as price, increase with substitution, so it remains for the papermaker to select the lowest cost product most fitted to his specific needs. Acetates and nitriles are used to a lesser extent and for essentially the same purposes as the ethylated. Cationic derivatives have gained acceptance as surface additives because of their specific property of electro-kinetic attachment to the negatively charged paper fiber. It is claimed that such products hold a high concentration of starch on the sheet surface and also remain attached to the fiber during the repulping process, thereby lowering the starch imparted BOD in the mill effluent (*27*). Manufacturers of fine paper grades who use cationic starch regularly for surface sizing claim cost reductions through lowered starch consumption, together with specific quality and operating improvements. These improvements include less TlO_2 usage; improved mullen, fold, and opacity; and reduced effluent loadings, both in solids and BOD.

In Plant Processing

As we have seen, the manufacturer supplies all grades of starch that are used as surface strength additives in essentially the same physical form as the native product, even though the molecular structure may be altered. It is extremely important that the molecules be fully dispersed from the granules for them to perform their required function. The first task of the papermaker, then, is to properly cook out the starch. The one exception to this requirement for initial dispersion occurs when starch is added uncooked to the furnish, or by spray, to be cooked out in situ in the dryers. In this event, the end result still depends on an ultimate complete dispersion. As previously noted, the papermaker also has three processes available that can be applied coincidently with cooking for reduction of viscosity—namely, high energy cookout, enzyme, and chemical conversion.

Atmospheric Cookout. Native starch at low concentrations, as well as most modified products and derivatives, needs only to be cooked prior to use. Cooking can be accomplished either by batch or by continuous methods. For batch cooking, concentration must be low enough that agitation can be maintained at the peak viscosity (gel point). The cooking temperature of 195°–200°F should be held for 20 minutes to ensure complete dispersion. The paste can then be diluted as desired and stored at a temperature of around 160°F for future use. Additives can be furnished either before or after cooking, depending on convenience and stability to heat, and the only requisites need be tankage, adequate agitation, and temperature indication and control. Direct steam injection is most widely used. Equipment for continuous cooking typically consists of the jet type cooker wherein starch slurry and steam are combined in an eductor type nozzle. To ensure complete dispersion, the cooking temperature at the nozzle should be maintained at 220° - 230°F by holding a slight back pressure on the outlet side. Such a cooker may be simple or may become quite elaborate, depending on the amount of controls and indicator-recording equipment that is desired.

High Energy Cookout. An adaptation of the jet cooker (*28*) provides the papermaker with a valuable means of viscosity reduction without depolymerization or impairment of the strength of the native starch molecule. The degree of viscosity reduction is limited, however, and is accomplished by applying an excess amount of steam and temperature to the starch at the cooking nozzle. It has been shown in the discussion of the pasting characteristics that the molecules within the granule are bound together as bundles or micelles. Under normal cooking conditions these bundles to a great extent remain intact and contribute materially to the inherent viscosity of the dispersed paste. The high turbulence or "shear" effects of the excess steam tear apart these micellar structures without undue breakdown of the molecules themselves. Viscosity is reduced to the level inherent in such a dispersion, and yet the strength of the long chain molecules is retained. From a practical standpoint the optimum benefits of this process can be attained by cooking at a concentration as high as 30% dry solids at a temperature of 175°C (350°F), with steam usage at 3 1/2 times the theoretical needed for cooking. Viscosity can be reduced to allow a manageable size press application at up to 8% solids, as compared with 4% for the parent unmodified product. For lower viscosities and higher solids application, the parent starch can be lightly modified, usually by oxidation, by the manufacturer. Alternatively, if a higher quality starch is required, a low fluidity derivative such as an ethylated product can be thinned by the high-shear process. In either event the full advantage of the higher strength molecules is maintained in the less highly chemically modified parent product.

Design of the high-shear cooker is similar to the jet type with the addition of steam flow control, a controlled back pressure valve, and much more sophisticated instrumentation. Operation can be completely automatic to maintain a ready supply of consistently uniform product for the paper machine or even to prepare alternately two different type starches. Automatic dilution systems can be provided to reduce the high solids to the concentrations at the point of use.

Enzyme Conversion. Enzyme conversion is the most commonly used in plant process for the thinning of starch for both size press and coating adhesives. The process is simple and economical and functions coincidently with the cooking stage. It does, however, require careful control to ensure a satisfactory quality end product. Slurried starch is heated and reacted with a biocatalyst by which both amylose and amylopectin molecules are progressively cleaved in a random manner to shorter chain lengths. When the viscosity is reduced to the desired point, the enzyme action is deactivated either chemically or by temperature. The alpha-amylase type enzymes are best suited for this type of conversion, since they initially liquefy the starch with a minimum of immediate degradation. Given sufficient time, however, alpha-amylase reduces the molecules to low molecular weight polysaccharides or "reducing sugars," which are of little value in adhesive properties and represent wasted material insofar as a strength additive is concerned (*29*). Under the best of conditions, enzyme conversion increases the proportion of short length molecules, and some reducing sugars are present in the final product. Basically, then, such product does not compare in quality with a supplier's prepared modified starch from which such short chain fractions are removed during manufacture.

Optimum conversion conditions depend on the specific enzyme used and are recommended by the manufacturer. Broadly speaking, best reaction conditions are in the range of pH 6.5 to 7.5 and temperature 70° - 75°C (160° - 165°F). Sodium, potassium, and calcium ions stabilize enzyme activity. Magnesium (other than as chloride) can be objectionable. Several metal ions are toxic. In fact, copper or tin salts are sometimes used as deactivating agents (*30*). Most enzymes deactivate at 95°C (over 200°F), although types that are stable up to this temperature are available. Required defoamers or bactericides must not be toxic to the enzyme, and starch should be low in protein and other non-starch impurities. Grades sold for enzyme conversion are usually pH adjusted by the manufacturer to fall within the proper pH range when slurried in the customer's process water. A calcium salt may be added if the customer's water is soft, though normal water hardness usually provides sufficient calcium stability. Too high a proportion of calcium ions can inhibit the deactivation step at the end of the conversion cycle with the result that enzyme reaction may continue in storage while awaiting use on the paper machine.

It is generally agreed that starch must be gelatinized to obtain liquefaction by enzymes. For this reason, any conversion cycle must go through the initial peak viscosity stage before the thinning begins to take effect. This can be a problem at the high concentrations used for preparing coating binders. Converting for size press use is usually carried out at lower solids, usually only enough higher than that used on the machine to allow some dilution water for cooling. Schwalbe (*31*) notes a peak viscosity of 35,000 cp (Brookfield) for 2 to 3 minutes at 15% concentration. These high ranges can be reduced by a slight modification by the manufacturer; however, this step adds an additional cost factor to the process.

Conversions can be carried out by any of three general time-temperature programs, depending on the capabilities of the particular equipment and the rate of usage by the paper machine. In the usual method, the starch slurry is thoroughly dispersed with the enzyme in the proper proportion, is heated through the gel stage to the optimum conversion temperature, and is held for a 20 to 30 minute reaction period. Then it is heated to 95°C for 10 to 15 minutes for the kill. The quantity of enzyme is adjusted to give the required viscosity in the finished product. Another method calls for an increased amount of enzyme and a "straight through" cook without a holding period. This method requires a carefully controlled rate of temperature rise to ensure a consistent reaction time. The third method, using a two-stage conversion, can be applied to high concentration pastes. In the initial step the temperature is raised to approximately 70°C (155° - 160°F) and held for a period to break down the initial gel. Heating is then continued to 75°C (167°F) for the final hold period (*32*).

The heart of any conversion system lies in reliable temperature-time control for ensuring consistent treatment from batch to batch. Either a cam controller or continuous viscosity control may be employed for this purpose. Although direct steam is usually used, a combination of direct steam with a steam jacket may be of advantage, the latter being used to maintain temperature during the holding period. Direct steam introduces condensate into the cook, and this must be taken into account when calculating finished concentration. As an added caution, use of a steam jacket must be carefully controlled. Residual heat in the jacket when the tank is low or empty forms a skin of dried starch gel which contaminates the subsequent batch. Agitating equipment must be rugged and must be designed to keep the starch paste moving during the peak viscosity period to prevent formation of pockets of nonuniform temperature and inconsistent conversion rates. Processes and equipment have been developed for continuous enzyme conversion and for combining enzyme conversion with high energy thinning, either on a batch or a continuous basis (*33,34,35,36*). In the latter instances, again, full advantage is taken of the dispersing action of high energy forces, with enzyme conversion only serving to reduce viscosity beyond their scope of effectiveness. Enzymes that are heat stable up to 100°C are available, and this enables initial continuous cooking of the paste by an atmospheric jet unit, followed by a reaction period for the enzyme to function and a final combination of enzyme kill and high energy thinning through a high shear cooker.

Chemical Conversion. Processes and equipment have also been described for chemical conversion of starch by such oxidizing agents as hydrogen peroxide and ammonium persulfate with suitable catalysts (*37,38,39,40*). These processes are aimed toward the same advantages of enzyme conversion—namely, the in plant, low cost conversion of a base-priced starch. It is also claimed that lower quantities of degradation products are produced. Again, the processes may be batch, continuous using a holding chamber for reaction time, or combined with high energy thinning. As with enzyme conversion, close process control is necessary to ensure a uniform product.

The three types of in plant prepared products offer the most economical additives for the papermaker, since they all have the advantage of a base price or at least a low-differential priced parent product. The importance of proper control of the enzyme or chemical conversions cannot be over emphasized. A degradation in the quality of the starch can easily result in rejections of off-quality paper. Since starch costs comprise only a small portion of the total costs of paper manufacture, it is axiomatic that the savings gained by purchase of a low price product versus a premium grade can be quickly wiped out by the losses through rejection of only a small amount of finished paper. These types of starches are used in the greatest volumes over the full spectrum of paper grades requiring size press treatments.

General Applications

Thus far discussion has been confined to the products and processes generally used to prepare the surface formulations for strength improvement, concentrating on corn starch as the most important in terms of volume used. Some of the considerations affecting the choice among the several types of starch that are available follow.

Retrogradation and Impurities. The formation of crystalline retrogradation precipitates mentioned earlier (page 40) can be a function of the type of modified starch used, as well as the method of preparation in the mill. This condition, more familiarly known as "amylose formation," can be a major problem to size press application, since the crystalline granules agglomerate into particles large enough to resist passing through the size press nip. They can accumulate in the supply system to form a heavy sludge and cause an extreme buildup in paste viscosity. Those that do pass through the nip contribute to excessive scaling and dusting from the sheet surface. Under favorable conditions, amylose precipitate can form from almost any of the commercial starch suspensions; however, the potential for trouble depends upon the relative proportion of straight chain molecular groups present. Thus, any of the starches containing the usual 25% to 30% amylose can form the precipitate. Acid modification cuts the amylopectin molecules in such a manner as to leave a larger proportion of straight chain fragments, so that this type of starch is more prone to precipitation. On the other hand, oxidation reacts to attach functional (carboxyl) groups which inhibit reassociation. For this reason pastes of oxidized starch have much better stability and less of a tendency for crystallization than those of "thin boils" (*41,42*). Such derivatives as ethylated products are quite stable against amylose formation, the theory being that the molecular groups attached to the straight chain molecules inhibit or make impossible the retrogradation caused by hydrogen bonding.

The manner in which the starch is cooked and handled assumes importance equal to that of the choice of starch in the consideration of amylose formation. Both the high energy and chemical conversion methods can be particularly troublesome under certain methods of operation. This can be most easily understood when one considers processes for the deliberate separation of the amylose and amylopectin fractions by crystallization of the former. By one process (*43*), starch at up to 10% dry solids is cooked at elevated pressure, approximately 150°C (300°F), and pH near neutral. This is done most conveniently by high pressure jet similar to the commercial high energy cooker. After cooking, the paste is rapidly lowered to atmospheric boiling point and is then slowly cooled over a period of up to 40 hours to around 30°C (85°F), during which time the crystals of amylose grow and settle from solution. It is evident that these conditions could be easily duplicated in a system for preparation of a size press furnish. Likewise, certain conditions of chemical conversion have the effects of acid modification and produce a product prone to amylose formation.

Amylose formation can be prevented by avoiding any of the favorable conditions noted above—for example, the slow cooling during storage. The presence of seeding crystals speeds up amylose formation so that a scrupulously clean system must be maintained.

Amylose can be detected by both solids measurement and microscopic examination, but amylose solids cannot be detected by refractometer. Amylose can be suspected, then, if a gravimetric solids measurement is appreciably greater than the refractometer solids. An amylose precipitate can be screened out by 325 mesh screen. The granules may appear by microscopic examination to be similar to a raw starch granule, even to the extent of showing birefringence. They can be differentiated, however, through their property of resisting cookout on boiling. More often, the crystals agglomerate into characteristic dumbbell, popcorn ball, needle, or platelet formations which soon become familiar to those working with these precipitates.

The importance of efficient refining of starch to good viscosity stability has been mentioned. Protein is an impurity of starch which can be troublesome to size press operation if present in excessive quantities. As with amylose, large size protein particles are subject to nip rejection and build up in the supply system to form an objectionable yellow to gray scum or sludge. Excessive oil residuals in the starch also contribute to this problem, since, if they are hydrolized to fatty acids, on processing they tend to react with the calcium and magnesium salts present in hard water to form fatty soaps. A typical refined starch contains approximately 0.35% protein and 0.04% residual fat (7).

Environmental Considerations. From the environmental aspect, starches offer two problems that would probably not be encountered with a synthetic type additive. Since starches represent the principal dry strength surface additives, these problems must be recognized. Herrick (*44*), Wilhelm (*45*), and Gillespie et al. (*46*) offer comprehensive information regarding the problems related to stream pollution. Surface applications offer the highest degree of sheet retention, with primary losses to the sewer being confined to minor spills and cleanups. Unfortunately, surface starch becomes a factor in pollution by its presence in broke added to the machine furnish. The magnitude of this problem can be quite independent of the starch actually being used in the mill, if substantial quantities of reclaimed starch treated broke from outside sources are used.

Of the many types of modified and derivatized starches, the oxidized types show the greatest effects toward dispersion of pigments. As little as 0.1% present in the furnish has been reported to affect pigment retention. This dispersion effect is thought to be a protective colloid action enhanced by the carboxyl groups attached to the molecules during the oxidation reaction. It is objectionable not only because of pigment losses from the sheet but also because of increased costs of effluent treatment. High pigment suspension prevents the reuse of the white water in many points on the machine, and the resultant increased fresh water volume only aggravates the problem. Finally, the high turbidity of the effluent which can result from even a relatively low level of pigment is very objectionable from both aesthetic and environmental standpoints. While it may be possible to successfully treat the white water, the more economical solution calls for an attack at the source. One step which has been found to be effective is the separate beating of the broke before addition to the furnish, which tends to minimize the dispersive effect. Another measure which is much more positive in controlling machine broke is the use of starch derivatives in place of the oxidized ones. Herrick (*44*) has reported an 84%

reduction in effluent turbidity during a three month controlled mill run substituting an ethylated for an oxidized starch. Cationic starches are also reported to be very low in dispersing effects.

The other problem inherent in practically all starches is the product leached from the fibers during the repulping of broke. This contributes directly to the BOD level of the effluent, which ranges from 0.5 to 0.75 pound BOD per pound starch (*46*). Use of more efficient derivatives, such as ethylated products, reduces the amount of necessary add-on, thereby reducing the starch going back with the broke. In addition, cationic surface size has been mentioned (page 43) as an alleviant for starch contributed BOD.

Viscosity Considerations. Viscosity has often been cited as a property of a given grade of starch that defines its usefulness to the papermaker. The importance of close viscosity control to the satisfactory performance of the product on the paper machine will be demonstrated subsequently. For all of its importance, a positive viscosity value cannot be tested and assigned to a given starch paste without including the history of the sample, method of preparation, and conditions of the tests. Viscosity, also referred to as consistency, can be simply defined as the resistance of a liquid to stirring (shear); fluidity is its reciprocal. The viscosity of a cooked paste depends on the composite effect of at least three known factors—namely, the content of swollen, elastic granules and granule fragments, molecular viscosity of the high-polymer molecules, and the structural network set up by reassociation. The viscometer can only measure the inertia of the starch paste—that is, its resistance to flow or distortion. It cannot differentiate between true viscosity and the effects of these three factors (*47*). Viscosity, then, in addition to being characteristic of the source species and type of primary treatment, depends upon such conditions as starch solids concentration, degree of agitation, time cycle and temperature during cookout, temperature changes after cooking, conditions of storage, age, and pH. Pastes of two species, such as corn and potato, may markedly differ in appearance and handling characteristics and yet measure the same in viscosity by a given test method. On the other hand, different methods may respond differently to a given paste. As an example, comparative viscosity measurements on pastes made from a low fluidity "thin boil" and a mid-range ethylated product are noted. Viscosities by Brookfield (rotating spindle) tested 6200 and 3200 centipoises, respectively. Tests by No. 2 Zahn Cup (orifice) showed just the opposite relationship, 25 and 52 seconds, respectively. Finally, viscosity changed with rate of shear, as indicated by Brookfield tests run at different speeds. The viscosity may drop with increased speed, indicating a thixotropic condition, or may increase, indicating dilatancy.

The many methods of testing that are used in the starch and paper industries usually employ either of two basic principles, resistance to a rotational spindle or flow through an orifice. More important examples of the former are the

Brookfield, C. I. V., Brabender, and Stormer viscometers; and examples of the latter include the Dudley pipet and Zahn Cup. References previously cited (*3,4,5,8*) include comprehensive discussions of the theories of viscosity development and methods of measurement. This section is limited to points related to viscosity control within the paper mill.

Orifice measurements are favored for viscosity control at the point of use, since the instruments are simple and inexpensive and tests can be easily run by operating people. Of the rotational instruments, the Brookfield finds most favor. Tests can be run quickly; however, the procedures are not as well adapted to routine testing. Brabender and C. I. V. are little used in paper mills. All tests have their limitations; thus, several points of caution are worthy of note. As noted above, the variables in preparation and handling of the sample are extremely critical for reproducible results. Too much importance cannot be placed upon cleanliness of equipment and strict following of standard test procedures. Temperature in particular must be carefully controlled, since it represents a constantly changing condition throughout the duration of any test. Viscosity tests falling near the ends of the range of the apparatus used can give misleading results. As an example, the Dudley pipet is a 100 ml glass pipet with a tip calibrated to pass that volume of water at 100°F in 34 seconds (*48*). The rate of delivery of a starch paste is a function not only of viscosity but also of drag within the stems and on the walls of the main chamber. In this respect, delivery times of a starch differ, depending on the physical dimensions of the pipet (which may vary among suppliers), even though water times may be the same. Use of a reference pipet to calibrate each new one avoids a problem from this standpoint. Dudley readings of over 60 seconds or so are questionable because of the temperature change during the period of the test and the excessive drag of the high viscosity paste. At the other extreme, Dudley readings of under 40 seconds indicate starch that is almost water thin. Qualities could swing through considerable percentage ranges without showing on the actual test value. In the case of an enzyme converted starch, Dudleys in the low range can indicate severe over conversion. From a practical standpoint, the effective Dudley range of 40 to 55 seconds nicely brackets the best operating range on most size presses. In the case of Brookfield measurements, the appropriate spindle should be chosen to keep the readings within the mid 80% of the scale.

Dudley or Zahn tests are used in many mills as routine periodic tests to monitor the quality of starch to the size press. Operators may easily be lulled into a sense of false confidence when tests over a long period of time appear to have little variation. Actually, there may be variables that profoundly affect the application that the orifice test may not pick up. A routine program of checking by a different method—for example, the Brookfield—can serve as a protection against such developments.

METHODS OF SURFACE APPLICATION

Four present day methods of applying additives after sheet formation have evolved during the long history of the papermaking process. The original method, which employed tub sizing, dates back to the earliest days of manufacture by hand sheet mold. Tub sizing was adapted to the continuous process after the advent of the fourdrinier machine and is still used for the special properties it imparts to fine paper grades. The size press is a natural development of the tub size process wherein the saturation tub is replaced by a pair of squeeze rolls. Today, by far the greatest proportion, probably in excess of 90%, of the surface additives used in the paper industry (exclusive of coatings) is applied at the size press. Calender sizing is limited to heavier weights of paper and board. Its primary purpose is to enhance surface qualities and is of less importance strictly to strength improvement. Finally, spray offers a convenient means of application that can be used not only on the fourdrinier after sheet formation but at other locations as well. Each of these methods has its advantages and limitations and is discussed in more detail in order of location in the papermaking process.

Spray Applications

Fourdrinier spray techniques utilizing uncooked starch slurry have been developed and ensure a convenient and trouble free means of gaining advantages toward strength improvement similar to a furnish addition or, to a lesser extent, a size press treatment. The advantages over, for example, a cooked starch furnish additive — or for that matter, a cooked starch spray — include elimination of the need for cooking equipment and better retention of the granular starch by the filtration properties of the formed sheet. The slurried starch at the point of application has no effect toward "slowing up" the sheet formation. Overall effects on operation are very similar to any filler product. Physical properties of unswollen corn starch granules, as compared with filler clay, are particle sizes of 5 to 25 microns (hexagonal granules) versus 0.5 to 10 microns (platelets) and specific gravity of 1.5 versus 2.58, respectively. It is interesting to note that the average corn starch granular size is half the width of a paper fiber and that approximately 200 million granules are present per gram, as compared to 3 million fibers per gram in softwood pulp (*49*). Any accumulations on the machine frame or surrounding area that result from a spray application can be easily washed down along with the normal stock buildups. Finally, as will be shown, there should be practically no contribution to effluent BOD. A spray application cannot be expected to match a full scale size press treatment. However, it can sufficiently improve the top side wax pick or linting qualities of a sheet to equal those of the wire side, which is of particular advantage to the production of low cost sheets for offset printing. Compared to a size press, equipment is relatively inexpensive; there is no limitation to machine capacity by need for additional drying; and application is quite safe to machine operation. A spray can be easily started and stopped as needed, and application rates can be conveniently controlled within wide limits (0.5% to 2% on machine output) without interruption to production.

Since an uncooked slurry provides the optimum advantages to be gained by a fourdrinier spray, the only required equipment includes conventional tankage, agitators, and pumps for slurry makeup and transfer. The spray boom is made up with suitable nozzles and is located over the wire at a position which gives the desired qualities, as noted below. Flat fan type nozzles have been found to give the best contact pattern on the wire and allow precise control of the point of application. Nozzles are spaced on the boom to give multiple overlap of the individual fans. A four-fold overlap has been found to be necessary for the most even coverage at high speeds, particularly in excess of 2000 fpm. The positioning of the boom over the wire depends primarily on the balance of internal to surface strength improvements that are required, although the presence of interfering cross walks, pipes, dandy, or other equipment and tendencies toward misting must be given consideration. A good starting point has been found to be a height of about 24 inches over the wire with the spray directed at about 30° to the horizontal against the travel of the wire. This allows about a 48 inch cast for the spray and gives a four-fold overlap using 50° nozzles spaced on 12 inch centers. Height and angle can then be adjusted as necessary for minimum misting. Accumulations and dripping onto the wire from cross equipment marks the sheet and so must be avoided.

Several special points in equipment design are important for trouble free operation of the system. A strainer of equivalent 50 to 70 mesh that can be cleaned without shutting down the spray must be located ahead of the spray boom to protect the nozzles from plugging. A continuous circulation back to storage of about one third of the total flow to the boom effectively prevents settling out in the line; in fact, circulation must be maintained in all parts of any slurry system to prevent plugups. In this respect a slurry bypass and a wash water flush should be provided on the spray boom itself for use when the shower has to be shut off temporarily. Finally, a positive displacement supply pump, which is adjusted and controlled by a valve on the return line, has been found to maintain the most uniform back pressure on the nozzles.

Application rate is a function of slurry concentration, operating pressure, and nozzle characteristics (design, size, and number). From the standpoint of practical operation, the spray boom should be designed to meet the requirements of machine width, speed, and production rate. Nozzle characteristics, then, are fixed. The system should be operated at a uniform pressure that is optimum to the type of nozzles that are used. Slurry concentration remains the single means of application control and one that can be varied up to at least 15% solids without operating problems. A good starting point for the design of a new system provides the capability for use of an approximately 6% concentration slurry to provide 1%, or 20 lb per ton (usually found to be adequate), at the production rate of the machine. The system then has the potential for more than double the treatment rate if necessary. It should be kept in mind that distribution on the wire is optimum at the lowest slurry concentrations. Further equipment refinements for long term operation include such features as automatic slurry make-up, continuous dilution from high concentration storage (up to 30%) to use concentration, or applications to multiple machines.

Spray contact at or near the dry line holds the maximum amount of starch on the top surface of the sheet; however, the coverage becomes splotchy and streaked if there is not enough surface water remaining to disperse the slurry droplets. A point several feet ahead of the dry line, depending on speed and stock freeness, has been found to give the best balance between evenness of coverage and surface improvement. Gains of three to five points in Dennison wax pick and up to 30% in IGT have been reported for this type of treatment (*50*). If there is no subsequent size press or coating, the distribution pattern can easily be checked on the dry sheet by painting or spraying with a dilute iodine solution. Moving the spray toward the headbox progressively allows more of the starch to work into the sheet to improve internal strength. In the writer's experience, the effects of such manipulation of position have given gains of up to 20% in Scott bond and of one to two Dennison wax points on the wire side. As the boom approaches the headbox, retention with resultant strength improvement decreases because of the poorer filtration efficiency of the lightly formed sheet. It should be noted that some starch passes through the wire at almost any location of the spray. Relative loss can be estimated by microscopic examination of the white water from a roll or suction box following the spray. Such presence of granular starch can be recovered and returned to the furnish via an efficient saveall system. Starch-contributed BOD in the final mill effluent is avoided, in contrast to either furnish addition or spray of cooked starch, up to 50% of which is irrecoverably lost (*51*).

The advantages of spraying a slurry versus a cooked out product have been noted. Complete cookout, however, remains fully important to the final efficiency of the starch toward strength improvement. Such cookout must take place in the early part of the dryer train while sufficient moisture is still present for complete hydration. It is important that the starch reach its gel point quickly, and for this reason a starch derivative with lowered gel point is of advantage on high speed machines (*52*). Ethylated products in the higher levels of commercial substitution perform very well in this function. Exceptions to this general concept are applications on slower speed machines or on machines where stock temperature is high enough to prematurely gel the starch to cause buildup in the wet presses or early dryer surfaces. The choice of product modification depends on machine speed, as well as specific requirements. Usually a well dispersed starch of the highest possible molecular length gives the best strength improvements. On slow fourdrinier and board machines, a high viscosity product has the time to flow among the fibers after cookout. At high speeds, a highly modified, high fluidity product may be necessary for good dispersion, even though there may be some sacrifice in basic film strength through the greater degree of depolymerization. The degree of dispersion and cookout can be checked by microscopic examination of iodine-stained samples of the finished paper, providing there is no size press or coating treatment. It is reported that cooked out cationic starch can be sprayed and can be retained on the fiber. However, the advantages of avoiding cooking and promoting efficiency in general housekeeping are lost.

The preceding discussion has been confined to use on fourdrinier machines. Spray booms can be set up on cylinder machines to spray between specific plies to improve ply bond at required locations within the board. For such use, a boom made up of flat sprays spaced for single coverage is mounted on the machine frame to direct the spray into the nip of the upcoming cylinder mold and the overhead sheet. Owing to the slower speeds and high temperatures, starch products of lower modification and higher viscosities are usually found to be adequate.

Tub Sizing

Tub sizing is worthy of brief discussion, even though the process is limited to only very high quality fine papers, which comprise only a very small proportion of present day production. Strips of Egyptian papyrus cemented together with starchy adhesive have been found dating back to 3000 B. C. The first recorded accounts of the use of starch as a size were set down by Pliny the Elder, who lived from A. D. 23 to 74, and described documents of 130 B. C. which were made by sizing papyrus with a modified wheat starch produced from fine flour boiled with vinegar. The Chinese are credited with first using starch size on true paper as early as A. D. 312 (53). If one can envision that such applications were made by dipping the handmade sheets into containers of prepared size, it can be seen that "tub sizing" as a process is as old or older than papermaking itself. In present day handmade paper operations, an "older process" for sizing is described as simply the immersion of stacks of 100 to 150 sheets in a copper tank of hot size, followed by light pressing to remove the excess. A "more modern method" is described by which the sheets are carried singly between endless felts through a vat of size followed by a pair of press rolls (54). Similar equipment adapted to handle the continuous web from the fourdrinier can be located within or at the end of the paper machine dryer section. The partially dried sheet is passed through a long pan and is followed by a squeeze arrangement to remove excess fluid. Drying can be at low tension over after-dryers or by leading off in loops through which warm air is circulated. In some cases the sheet may be reeled up in a "wet pack" for a short storage time before drying to allow distribution of the size (55).

Animal glue is largely used for tub sizing, both alone and in combination with prepared starches. Low viscosities are necessary to achieve necessary penetration into the sheet. Starches may be modified by oxidation, acid treatment, enzyme conversion, or dextrinization. The process is quite sensitive to retrogradation and formation of amylose precipitates, particularly when glue is part of the formula (*56*).

Tub sizing represents the costliest means of surface application, because of the relatively expensive materials used and the limitation of production rates. Such costs are justified by the special qualities that can be imparted to high quality papers—for example, rag bonds, map papers, and chart papers. The longer contact with the sizing medium results in up to 25% more pickup than is accomplished by size press (*57*). In addition to special embossed and cockle finishes, superior strength qualities, particularly fold endurance and erasability, can be imparted to the sheet. The tough surface film plus the adhesive qualities of the starch taken up by the sheet improve the latter quality in two ways—first, by reducing penetration of ink into the paper, making it easier to remove by erasure, and, second, by increasing fiber bonding and thus increasing resistance to abrasion (*58*).

Size Press

The size press is an outgrowth of the tub sizing operation, whereby a pair of squeeze rolls replaces the size tub and starch is applied directly to the rolls. The term "tub sizing" persists and is still often used in reference to modern size press operations. As previously noted, the great volume of the presently used surface additives exclusive of coatings are applied at the size press. Normal pickup rates cover the range of 2% to greater than 5% solids on sheet weight. This compares with effective 0.5% to 1.5% by furnish addition, 0.5% to 2.0% by spray, greater than 5% by tub sizing, and a maximum of 2% by calender. Thus, it can be seen that the size press is the most versatile of applications in terms of the amount of material that can be incorporated onto the sheet. The major drawback of the size press operation is the large reduction in machine production capacity because of the need to dry the moisture from the applied starch. This factor assumes importance when the requirements for added strength or surface qualities are marginal; for example, for lint control on an offset sheet, a choice may be made between a size press and treatment by either furnish addition or spray.

Quite simply, the size press is made up of two press rolls which may be mounted one above the other (vertical), side by side (horizontal), or, in a more modern design, with the top roll inclined at about 45° toward the dry end of the machine. One roll, usually the top, is a hard roll, while the other is covered with rubber of hardness in the range of 20 to 30 Pusey and Jones plastometer. The bottom roll of the pair is driven by the main paper machine drive, while the top is usually driven by a speed controlled helper drive. On smaller installations the top roll may ride free, although this is undesirable since the drag on the sheet in turning the roll affects the pattern of size pickup. The amount of pickup depends to some extent on the softness of the rubber and the width of the nip, so to assure uniform treatment it is important that the rolls be maintained in good condition and be properly crowned to give a uniform cross-machine nip pattern. A spring loaded paper roll is located ahead of the press to control the lead in angle of the sheet. The sheet gains in width as it takes up size in the press nip and shrinks rather rapidly in the initial stages of drying. A spreader bar or Mount Hope roll is usually located after the press to spread the sheet to avoid wrinkles, and the first one or two after-dryers are usually run without dryer felts to allow the sheet to stabilize without restraint. Ideally, the location of the size press within the dryer train is at the spot where, under average operating conditions, the sheet can be dried to the proper moisture level going into the press and sufficient after-dryers are available to dry out the remoistened sheet to the finished moisture level. Under these conditions the machine is not "speed limited" by inadequate after-dryer capacity. Such ideal conditions are rather rare, since many day-to-day variables affect the after-drying requirements, a few of which are sheet basis weight and absorption characteristics, starch characteristics, and machine speed.

In the vertical press, the sheet is led into the nip in a downward direction, typically at about 25° to the horizontal. Starch solution introduced by shower to the top side forms a puddle on the top of the sheet, where the area depends on the angle of sheet entry. Excess starch spills off the edges at the nip, so that the actual pond width varies to some extent from center to edge. This volume of starch allows a short but appreciable time for absorption into the sheet ahead of the nip. On the other hand, starch showered onto the bottom roll is carried directly into the nip with little opportunity for preabsorption. These conditions result in some nonuniformity from center to edge on the top side, as well as some two-sidedness of starch takeup, which may or may not be accentuated by the normal differences in top to wire side inherent in sheet formation. Another disadvantage is the weight of the pond, which may tend to deform a sheet of light basis weight. The horizontal size press was designed to overcome these shortcomings and was followed by the inclined press which helps avoid the awkward vertical lead-in of the sheet. Starch is showered onto the nip on each side of the sheet to provide equal conditions of contact.

The factors affecting the pickup of starch at the size press are several and interrelated, and some of the net results are quite controversial throughout the industry. To cite one example, it is stated (59) that "A survey in the paper industry of the effect of nip pressure on the weight added to raw stock gave 50 answers of increased weight by higher pressure, with 50 operators disagreeing" (59). A view of some of the generally accepted relationships, together with some of the results reported by several investigators who have studied both pilot and commercial machine treatments, follows. However, one must bear in mind the fact of general controversy and realize that each situation of application is unique within itself.

The important factors in press operation are summarized (*60*) as sheet properties of basis weight, density, smoothness, internal sizing, and moisture; press roll characteristics of hardness, crown, and loading; machine configurations such as angle of entry, degree of wrap, and release from the press; and operating factors such as depth of pond, length and tension of draws, and nip pressure. Properties of the size solution itself, including solids, viscosity, and temperature, comprise a most important additional category. The generally accepted effects of some of these factors by themselves can be briefly stated. Optimum moisture content of the oncoming sheet, in the general opinion of operators, is in the 4% to 12% range (*61*), preferably on the high side, although moisture too high gives operating problems. Size pickup is reduced at moistures below 4%, with the starch being held as a surface film with little penetration. Surface cracking, dusting or scaling, and buildups on the after-dryers can result from this condition. It is common to find mills over-drying the sheet to the size press either to correct poor moisture profiles or to compensate for inadequate after-dryer capacity. An even moisture profile is necessary for uniform cross machine absorption; however, correction should be by some means other than over-drying. Size pickup is reduced with increase in internal sizing. Usually a ratio which depends on the desired properties of the sheet is maintained. It is extremely important that internal sizing remain uniform, since the external treatment is very sensitive to changes. Since pickup and penetration (short of complete saturation) are practically independent of sheet weight, the percentage of pickup to total sheet weight decreases with increased basis weight. Size absorption is also reduced with increased sheet density and/or smoothness, other factors being constant. Viscosity is the most important property of the size solution itself, since penetration varies inversely as the square root of the viscosity. Solids concentration and temperature are important in their effects on viscosity (*62*). Concentration is additionally important in its own right by having a directly proportionate effect on the dry solids pickup. For this reason, if sheet qualities affected by surface sizing are not adequate, the first thing the papermaker usually does is to increase the concentration of the size. This may be self defeating if the reduced penetration (because of higher viscosity) aggravates the problem or if the sheet becomes harder to dry. Temperature has a practical upper limit of 150° to 160°F (65° - 70°C) because of the safety problems in handling and the limitation of the rubber roll coverings. With respect to press roll characteristics and loadings, the higher the nip pressure, the less the amount of size picked up, since the size is more effectively expressed from the sheet. In fact, it is stated that no other factor has an effect on pickup rate approaching that of a nip pressure change (*62*). However, such adjustments must be made with care to avoid too great a change in nip configuration. Roll hardness and diameter are important because of their effects on nip pressure. Softer and larger diameter rolls give nips of greater area resulting in less pressure per square inch at a given unit per lineal inch loadings.

In the choice of the type of starch to be used, the papermaker has a wide range of cooked paste viscosity-solids relationships available by which he can obtain the qualities that he desires. As a rule, the higher the inherent viscosity of the starch, the greater the strength improvement per unit of starch applied. On the other hand, the higher the fluidity of the starch (more modification), the greater the concentration at which the starch can be used and the greater the amount that can be added to the paper (*63*). A high viscosity unmodified or slightly modified starch can be used at low solids for a light surface treatment to lay fiber or to improve scuff resistance. Lower viscosity modified products allow higher solids pastes at a given cooked out viscosity, which in turn give a heavier starch treatment that still remains much on the surface to give such effects as improved wax pick or IGT, along with some boost to internal strength. Finally, internal bond, burst, and tensile can be maximized by using a medium to low viscosity starch applied at conditions that ensure maximum penetration. Tear is generally reduced under such conditions.

A brief view of the results of some specific studies of size press applications follows. These works make apparent the varied and complex interrelationships among the factors affecting size press pickup. A detailed study of the articles themselves should make the reader aware of the pitfalls that are encountered when these interrelationships are not taken into account when working with a specific application on a specific machine.

Hoyland and Howarth (*64*) report the development of a theoretical equation for size press wet pickup which was subsequently checked out on a commercial paper machine. Wet pickup is reported to be related to three factors—the "sheet term," the "absorption term," and the "hydronamic term," represented respectively by the three factors on the right hand side of the following equation:

$$P_u = K_1 + \frac{K_2}{(\eta s)^n} + K_3 (\eta s)$$

where:

P_u is the *wet* pickup by the press
K_1, K_2, K_3 are coefficients
η (Eta) is the viscosity of the sizing solution
s is the speed of the web through the size press
n is an index.

Coefficient K_1 for the sheet term is reported to depend only on the degree of beating of the stock for the base sheet, for which permeability is a sensitive measure. Additionally, this term is reported to control most of the pickup up to speeds of approximately 150 m/min (approximately 500 fpm). Coefficient K_2, the absorption term, was found to be dependent upon both solution viscosity and degree of beating of the sheet. The value of K_2 is specific to a given size press and base sheet.

The hydronamic term was found to assume greater importance, thereby making a steadily rising contribution to size pickup with increasing speeds. The theoretical work covered by this article was concentrated on this term. The coefficient K_3 depends on the film splitting factor, the roll radius and hardness (deformation), and nip loading. It is reported that for hard rolls (P & J under 30) the rising portion of the speed-pickup curve is straight; however, for softer rolls the rate of increase of pickup tapers off at higher speeds.

Contribution to wet pickup by each of the terms was checked satisfactorily on a commercial machine size press operating in the range of 14 to 76 cps viscosity and 180 to 210 m/min (600 - 690 fpm) speed.

Dill (65) reports investigations on two pilot machines that were made "toward the goal of aiding the man responsible for grade development and machine operation."

> "The increased usage of size press applications in the paper making process demands a more thorough understanding of related technology. Specific information on control and understanding of size press pick-up was the objective of this study. Internal sizing is the most significant factor in pick-up control but other variables are important. Variables included in this pilot machine study were machine speed (100-3000 ft./min.), basesheet porosity and sizing level, and starch concentration and viscosity. Statistical analysis of the data provides simple equations which correlate pick-up to the variables studied. A discussion relating the principal mechanisms of size press pick-up (capillary penetration, hydraulic penetration, metered by surface roughness, and hydrodynamic metering) further aids in understanding and predictability" *(65)*.

Two conclusions are especially worthy of note. Internal sizing was found to affect size press pickup even at very low levels which are undetectable by ordinary size tests. This suggests the possibility of significant changes in pickup through variations in sizing introduced by broke or recycle stock. Machine speed influences pickup through multiple avenues, while the effect of size solution viscosity is dependent upon base stock characteristics. As a result, either of these factors may increase or decrease size press pickup, depending on the conditions.

Chilson and Fahey (66) report a study involving applications of low and medium viscosity ethylated starches, as well as an enzyme converted starch on a horizontal size press. Pickup of all starches was found to increase with incoming sheet moisture over a range from 5 to about 15 to 20%. For the high viscosity starches (above 50 seconds Dudley), pickup continued to increase up to 35% moisture. At 5% moisture most or all starch remained in the sheet surface, while at 15 to 20% the sheet was completely penetrated (true of both rosin free and rosin sized grades). An unsized web was two to three times more receptive to starch than one with 1% internal size. The relationship of starch concentration to the amount applied to the web was nearly linear. Starch treatment improved all strength properties except tearing resistance, which dropped as depth of penetration increased. It was found that the starch held on the surface in the low-moisture application was just as effective toward burst and tensile as the starch that had penetrated the sheet, even at lower levels of application. The greatest strength improvement gained by the starch treatments was noted in surface pick resistance.

Studies along similar lines are reported by Walker (*67*), Solheim (*68*), Kirk et al. (*69*), and Dill et al. (*70*).

Gate Roll Size Press

The discussion thus far has related to the operation of the conventional size press, which is limited to the application of relatively low viscosity size formulations. The gate roll size press is a recent development introduced by the Beloit Corporation in the 1960's. This size press system employs the principle of metering and transfer rolls which has been successfully used on coating equipment for many years. An advantage of the gate roll size press is the ability to apply size formulations of much higher viscosities, which can be made up of less modified adhesives carrying higher inherent molecular strength. The installation consists basically of a conventional vertical size press with the size solution being metered onto either or both of the rolls by pairs of applicator (gate) rolls, rather than by shower. Application can be made either on one or both sides of the sheet and can employ either clear size or pigmented formulations.

Lettenberger (*71*) cites paper mill experiences with the gate roll size press. In addition to the factors attendant to conventional size press operation, control of the surface treatment can be further augmented by relative speeds and nip pressures of the gate rolls. Formulations with Brookfield viscosities of 1500 cps to 3000 cps (10 rpm, 140°F) are given as being optimum for proper operation, as compared with the range of around 50 cps for conventional presses.

Calender Application

The calender stack offers the final opportunity for surface treatment on the paper machine. Because of inherent operating problems, calender applications are confined to heavy weight papers and board grades. As with the size press, it is necessary to dry out the water portion of the size before reeling. A heavy sheet may carry enough heat to achieve this purpose, although in most cases a second "dry" stack with steam heated rolls or even additional dryers are necessary to maintain production rates. In any event, the overall volume of size that can be incorporated is limited to the drying capabilities of the overall system.

The effects most commonly sought in calender sizing have been listed (*72*) as changes in sheet absorption (water, oil, air, aqueous and non-aqueous mixtures), sheet structure (curl, fuzz, hardness, density, strength), surface chemistry (solubility, adhesive qualities, specific reactions), optical properties (color, gloss), moisture balance, and uniformity. Only the second category, strength, both internal and surface, is considered here. However, it must be remembered that all of these effects are interrelated to some degree, and no treatment is specific to one effect alone.

Again, starch is the principal sizing agent used at the calender stack and is the one having the greatest effect on strength. Depending on the type of product used, good results toward surface properties can be realized as well. Other additives, such as CMC, alginates, insolubilizing agents, lubricants, wax sizes, plasticizing agents, and non-skid agents, are generally used for their effects on surface properties only; and from a strength standpoint their value is limited to such qualities as scuff resistance, reduction in fuzz, and improved wax pick and IGT. The viscosity range of the sizing solution is limited by the mechanics of the particular operation, and starches over the full range of modification can be used at solids levels consistent with such limitations. As an example, a solution of unmodified (native) starch may be used at 2 to 5% solids to give a light surface film in order to obtain scuff resistance at minimum cost. At the other extreme, highly modified derivatives may be used at up to 24% solids to achieve the maximum surface qualities plus absorption for improved internal bond. (*73*).

Since a detailed discussion of calender construction and operation is beyond the scope of this chapter, the techniques of the sizing treatment itself are concentrated upon. The starch is prepared with necessary additives by methods previously discussed. The solution is applied in calender boxes, which are flat pans, six to eight inches in width, extending across the machine and mounted with a flexible lip against an upcoming calender roll leading into the nip. Dams are provided in each end of the box to confine the starch within the limits of the sheet width. The size solution is introduced into one or more inlets of the box, depending on machine width, and continuously overflows back to the supply tank through raised outlets so adjusted that a shallow pond is maintained in the box somewhat below the level of the roll nip. The starch flows onto the surface of the upturning calender roll and is carried into contact with the sheet in the nip. The rate of pickup by the sheet depends on the moisture content of the paper, the temperature and concentration of the sizing solution, location of the boxes on the calender stack, and the number of boxes used. Boxes may be located on one or both sides of the stack, depending on the surfaces to be treated. In addition, two boxes may be located in successive positions to provide a double treatment to one side. In this arrangement, the first box fills the pores of the paper and gives primary effect to surface strength, while the second application enhances the surface film. The boxes are usually located as high in the stack as possible. The oncoming sheet at the top of the stack is the most open and becomes more dense as it travels downward through successive nips. It follows, then, that the higher the box, the greater is the absorption and volume pickup. More moisture must be dried out; however, there is more time available for doing this and more opportunities for improving the finish. Care must be taken that the starch does not completely penetrate the sheet, since it can stick to the roll and break down the operation. Thus, the height that a box can be used may depend on the weight and porosity of the sheet, as well as the conditions of internal sizing. A tacky or high solids size may tend to lift fibers and build up scabs on the roll surfaces. Small amounts of lubricants, such as soaps, waxes, or glycols, are often used to control these problems. Sizing temperatures may range from 140° to 180°F. Normally the size is maintained by the temperature of the oncoming sheet and calender rolls. If the incoming flow from storage is materially cooler, local shrinkage of the roll opposite the inlet point may lead to caliper variations in the sheet. The success of a calender application depends on a uniform incoming sheet and calender rolls of uniform diameter. Cross machine variations in moisture, basis weight or caliper, aggravated by uneven rolls, result in nonuniform starch pickup with resulting unacceptable variations in finished sheet quality.

Literature Cited

1. J. W. Swanson, *Am. Pap. Ind.*, 52(3): 39-45 (1970).
2. J. P. Casey, "Pulp and Paper," 2d ed., Interscience, New York, 1960, vol. 2, pp. 946-1153.
3. R. W. Kerr, ed., "Chemistry and Industry of Starch," 2d ed., Academic Press, New York, 1950, chap. 1, 3, 6, 7, 9, 11, 16, 18.
4. R. L. Whistler and E. F. Paschall, eds., "Starch: Chemistry and Technology," Academic Press, New York, 1965, vol. 1, chap. 1, 7, 10, 12, 14, 16.
5. R. L. Whistler and E. F. Paschall, eds., "Starch: Chemistry and Technology," Academic Press, New York, 1967, vol. 2, chap. 6, 9, 10, 13, 15, 17, 20, 25.
6. R. G. Macdonald and J. N. Franklin, eds., "Pulp and Paper Manufacture," 2d ed., McGraw Hill, New York, 1969, vol. 3, pp. 1-72, 576-627.
7. "Corn Starch," Corn Industries Research Foundation, Washington, D.C., et al., 1964.
8. F. H. Frost et al., "Starch and Starch Products in Paper Coating," TAPPI Monograph Series No. 17, 1957, pp. 3-107.
9. J. P. Casey, "Pulp and Paper," 2d ed., Interscience, New York, 1960, vol. 2, pp. 1104-1105.
10. J. P. Casey, "Pulp and Paper," 2d ed., Interscience, New York, 1960, vol. 2, p. 1136.
11. H. W. Leach, "Starch: Chemistry and Technology," ed. by R. L. Whistler and E. F. Paschall, Academic Press, New York, 1965, vol. 1, p. 292.
12. R. W. Kerr, "Starch and Starch Products in Paper Coating," TAPPI Monograph Series No. 17, 1957, p. 3.
13. J. C. Rankin, R. A. Webber, and P. J. Robbins, *Tappi*, 53(8): 1482-85 (1970).
14. J. C. Rankin, A. J. Ernst, B. S. Phillips, B. T. Hofreiter, and W. M. Doane, *Tappi*, 58(1): 106-8 (1975).
15. G. T. Austin and J. E. Sullivan, *Tappi*, 51(11): 53A (1968).

16. R. W. Kerr, ed., "Chemistry and Industry of Starch," 2d ed., Academic Press, New York, 1950, pp. 5-6.
17. R. W. Kerr, ed., "Chemistry and Industry of Starch," 2d ed., Academic Press, New York, 1950, pp. 14-16.
18. R. W. Kerr, "Starch and Starch Products in Paper Coating," TAPPI Monograph Series No. 17, 1957, p. 6.
19. R. W. Kerr, ed., Chemistry and Industry of Starch," 2d ed., Academic Press, New York, 1950, pp. 9-10.
20. J. H. Pazur, "Starch: Chemistry and Technology," ed. by R. L. Whistler and E. F. Paschall, Academic Press, New York, 1965, vol. 1, pp. 137-8.
21. D. French, "Chemistry and Industry of Starch," ed. by R. W. Kerr, 2d ed., Academic Press, New York, 1950, pp. 165-7.
22. D. P. Langlois and J. A. Wagoner, "Starch: Chemistry and Technology," ed. by R. L. Whistler and E. F. Paschall, Academic Press, New York, 1967, vol. 2, pp. 465-67.
23. R. L. Whistler and E. F. Paschall, eds., "Starch: Chemistry and Technology," Academic Press, New York, 1965, vol. 1, pp. 343-45.
24. R. W. Kerr, "Starch and Starch Products in Paper Coating," TAPPI Monograph Series No. 17, 1957, p. 10.
25. H. J. Roberts, "Starch: Chemistry and Technology," ed. by R. L. Whistler and E. F. Paschall, Academic Press, New York, 1967, vol. 2, p. 294.
26. E. T. Hjermstad, "Starch: Chemistry and Technology," ed. by R. L. Whistler and E. F. Paschall, Academic Press, New York, 1967, vol. 2, pp. 424-25.
27. K. J. Bristol and G. H. Brown, *Pap. Trade J.*, 153(27): 42-4 (1969).
28. V. L. Winfrey and W. C. Black, U.S. Patent No. 3,133,836 (May 19, 1964).
29. J. E. Maryanski, *Pap. Ind.*, 46(3): 209-12 (1964).
30. H. C. Schwalbe and E. P. Gillan, "Starch and Starch Products in Paper Coating," TAPPI Monograph Series No. 17, 1957, p. 45.
31. H. C. Schwalbe, "Pulp and Paper Manufacture," ed. by R. G. Macdonald and J. N. Franklin, 2d ed., McGraw Hill, New York, 1969, vol. 3, p. 75.
32. J. P. Casey, "Pulp and Paper," 2d ed., Interscience, New York, 1960, vol. 2, p. 1129.
33. L. E. Cave and F. R. Adams, *Tappi*, 51(11): 109A (1968).
34. F. G. Ewing and H. R. Harrington, *Chem. Eng. Prog.*, 63(3): 65 (1967).
35. F. G. Ewing, J. L. Bidrawn, B. H. Landis, and H. R. Wheeler, Jr., U.S. Patent No. 3,371,018 (Feb. 27, 1968).
36. H. C. Schwalbe, U.S. Patent No. 3,450,549 (June 17, 1969).
37. H. O. Kauffman, P. H. Eggertsville, K. Margulies, and J. R. Ryan, U.S. Patent No. 2,276,984 (March 17, 1942).
38. H. O. Kauffman, P. H. Eggertsville, K. Margulies, and J. R. Ryan, U.S. Patent No. 2,307,684 (Jan. 5, 1943).
39. G. E. Lauterbach, U.S. Patent No. 3,211,564 (Oct. 12, 1965).
40. H. W. Maurer, U.S. Patent No. 3,475,215 (Oct. 28, 1969).
41. H. C. Schwalbe, "Pulp and Paper Manufacture," ed. by R. G. Macdonald and J. N. Franklin, 2d ed., McGraw Hill, New York, 1969, vol. 3, pp. 65-6.
42. B. L. Scallet and E. A. Sowell, "Starch: Chemistry and Technology," ed. by R. L. Whistler and E. F. Paschall, Academic Press, New York, 1967, vol. 2, p. 243.
43. D. P. Langlois and J. A. Wagoner, "Starch: Chemistry and Technology," ed. by R. L. Whistler and E. F. Paschall, Academic Press, New York, 1967, vol. 2, p. 482.
44. R. Herrick, *Tappi*, 49(11): 79A-81A (1966).
45. L. K. Wilhelm, Western Michigan University thesis, Jan. 1968.
46. W. J. Gillespie, C. A. Mazzola, and D. W. Marshall, *Pap. Trade J.*, 154(9): 29-32 (1970).
47. M. J. Mason and W. R. Fetzer, "Starch and Starch Products in Paper Coating," TAPPI Monograph Series No. 17, 1957, pp. 65-71.
48. M. J. Mason and W. R. Fetzer, "Starch and Starch Products in Paper Coating," TAPPI Monograph Series No., 17, 1957, p. 74.
49. H. C. Schwalbe, "Pulp and Paper Manufacture," ed. by R. G. Macdonald and J. N. Franklin, 2d ed., McGraw Hill, New York, 1969, vol. 3, pp. 22, 62.
50. W. H. Thomin, *Pap. Technol.*, 13(2): 120-26 (1972).

51. J. P. Casey, "Pulp and Paper," 2d ed., Interscience, New York, 1960, vol. 2, p. 959.
52. J. P. Casey, "Pulp and Paper," 2d ed., Interscience, New York, 1960, vol. 2, pp. 950-51.
53. R. L. Whistler, "Starch: Chemistry and Technology," ed. by R. L. Whistler and E. F. Paschall, Academic Press, New York, 1965, vol. 1, pp. 1-2.
54. R. G. Macdonald and J. N. Franklin, eds., "Pulp and Paper Manufacture," 2d ed., McGraw Hill, New York, 1969, vol. 3, pp. 10-11.
55. R. G. Macdonald and J. N. Franklin, eds., "Pulp and Paper Manufacture," 2d ed., McGraw Hill, New York, 1969, vol. 3, pp. 74-5.
56. R. W. Kerr, ed., "Chemistry and Industry of Starch," 2d ed., Academic Press, New York, 1950, p. 493.
57. J. P. Casey, "Pulp and Paper," 2d ed., Interscience, New York, 1960, vol. 2, p. 1118.
58. J. P. Casey, "Pulp and Paper," 2d ed., Interscience, New York, 1960, vol. 2, p. 1112.
59. H. C. Schwalbe, "Pulp and Paper Manufacture," ed. by R. G. Macdonald and J. N. Franklin, 2d ed., McGraw Hill, New York, 1969, vol. 3, p. 73.
60. J. P. Casey, "Pulp and Paper," 2d ed., Interscience, New York, 1960, vol. 2, p. 1116.
61. J. P. Casey, "Pulp and Paper," 2d ed., Interscience, New York, 1960, vol. 2, p. 1121.
62. H. Hardman and E. J. Cole, "Papermaking: Questions and Answers," Lockwood Publishing Co., New York, 1969, pp. 128-131.
63. J. P. Casey, "Pulp and Paper," 2d ed., Interscience, New York, 1960, vol. 2, p. 1130.
64. R. W. Hoyland and P. Howarth, *Pap. Technol.*, 13(1): 38-40 (1972).
65. D. R. Dill, "Control and Understanding of Size Press Pick-Up," Paper presented at the TAPPI Coating Conference, Miami Beach, FL, April 30 - May 2, 1973, pp. 143-153.
66. W. A. Chilson and D. J. Fahey, *Am. Pap. Ind.*, 48(3): 81 (1966).
67. W. P. Walker, *Pap. Technol.*, 7(1): 38-44 (1966).
68. P. F. Solheim, *Tappi*, 49(4): 82A-83A (1966).
69. D. C. Kirk and E. R. Baker, *South. Pulp Pap. Manuf.*, 30(8): 84 (1967).
70. D. R. Dill and P. R. Graham, *Tappi*, 50(3): 133-36 (1967).
71. J. E. Lettenberger, "Papermill Experience with a Gate-Roll Size Press," Paper presented at the TAPPI Coating Conference, Miami Beach, FL, April 30 - May 2, 1973, pp. 165-176.
72. J. P. Casey, "Pulp and Paper," 2d ed., Interscience, New York, 1960, vol. 2, p. 1119.
73. J. P. Casey, "Pulp and Paper," 2d ed., Interscience, New York, 1960, vol. 2, p. 1133.

3

Natural Gums as Dry Strength Additives

Fritz MK Werdouschegg

INTRODUCTION

A great many natural materials are used as dry strength additives in the manufacture of paper and board, both in the United States and abroad. Among these are gums and mucilages obtained from guar and locust beans, tamarind kernels, karaya, okra, and others. Of these, guar and locust bean gums have found the most widespread use; of the remaining ones, tamarind gum finds the greater usage.

This chapter deals specifically with the utilization of gums for paper or board strength development, the types of and handling of gums, and some specific uses of gums.

Special features of this chapter in the appendix section are the comprehensive, although not all-inclusive, bibliography of published papers relating to the use of or preparation of natural polymeric substances (specifically, gums and mucilages) and a listing of related domestic and foreign patents.

HISTORICAL BACKGROUND

The use of mucilages in the manufacture of various types of paper is an ancient tradition, dating back to the first papermakers in China. These mucilages were obtained from certain plants whose roots or stems were the source of the active natural polymeric substances. The mucilages were used, however, primarily as deflocculating and suspending agents for the relatively long fibered pulp.

As papermaking evolved from its primitive beginnings, it became more and more apparent that these mucilageneous additives provided the papermaker with more than just a convenient suspending agent. During the drying process of the paper, the polymeric material adhered to the fibers and held the wet fiber mat together. Thus, the first wet web bonding agent was recognized.

With the further development and evolution of chemical pulping, new fibrous materials became available to the papermaker. However, most of these fibers were substantially shorter in length than the customarily used grass and reed fibers of old. The new fibers required substantial input of energy for the development of fiber bonding properties. To alleviate this shortcoming, the papermakers investigated additives which they hoped would provide the necessary fiber-to-fiber bonding. Ever since, they have sought a means of "pouring hydration" out of a barrel. One of the first natural gum products used for this purpose came into use shortly before World War II; it was a mucilage prepared from the powdered endosperm of the locust bean (carob seed). At that time, locust bean powder had been imported and utilized on a limited scale for specialized uses in paper manufacture for which this somewhat expensive product could be justified (*1*).

Within a short period after the introduction of locust bean gum, the development of another source of gum began. The shortages of locust bean gum that were envisioned because of the conflict in Europe spurred researchers in the late 1930's to investigate the mucilageneous properties of the guar bean endosperm. The endosperm of both beans consists almost totally of a natural polymeric substance, galactomannan. However, it took several years before domestic cultivation became established. Furthermore, the processing of the guar bean presented milling problems of significant proportions (*2*). Nevertheless, the researchers' perseverance culminated in the development of a refining or milling process which successfully separated the gum-containing endosperm from the seed-hull and germ (embryo) (*3*).

The popularity of mucilages derived from both locust and guar beans has increased substantially. In the twenty years from 1953 to 1973, the total consumption of guar gum in the United States rose from less than 2.5 million pounds to well over 33 million pounds (*4*). By 1970, the world production of locust bean gum amounted to about 30 million pounds, of which only a portion found its way to the United States (*4*).

It was estimated that the combined consumption of guar and locust based gums in the paper industry amounted to about 17 million pounds by the end of 1969 (*5*). Since then, the steep increase in the price and relatively poor availability of locust bean gum has shifted the bulk of the consumption toward guar gum.

SOURCES AND COMPOSITION OF NATURAL GUMS

The most common gums and mucilages used in paper and board manufacturing are as follows (*6*): guar gum, locust bean gum, and tamarind gum.

TAPPI CA Report No. 60 lists a number of specific commercially tradenamed products under these headings:

1. Fiber Bonding,
2. Dry Strength, and
10. Formation.

Specific products within these categories are available from the following listed suppliers: American Key Products, Inc., Burtonite Co., Celanese Corp. (formerly Stein, Hall & Co., Inc.), Henkel Corp. (formerly General Mills Chemicals, Inc.), and Tragacanth Importing Co. (TIC).

In addition, there are other suppliers, usually importers or manufacturer's agents, who sell specific gum products.

Guar Gum

Guar gum is derived from the seed of the guar plant (Cyamopsis tetragonolobus) of the family Leguminosa. This plant is an annual pod-bearing, nitrogen-fixing legume. It is indigenous to India and Pakistan and was introduced to the United States in the early part of this century. Commercial production began on a large scale in the late 1940's and early 1950's (*2,7*).

The guar seed is composed of the following (*8*):

Endosperm	35-42%
Germ (Embryo)	43-47%
Seed-Hull	14-17%

From the above data, it is apparent that in order to obtain a relatively pure grade of galactomannan, the effective and desirable ingredient of the guar bean, the endosperm must be separated from the hull and germ. There are several processes which accomplish this (*3,7*). Depending upon the degree of refining, one obtains food grade, technical grade, or industrial grade guar gum.

The galactomannan of the guar endosperm is reported (*9*) to consist of a D-mannopyranosyl chain (backbone) to which two D-galactopyranosyl units (galactose to mannose ratio of 1:1.49) are attached for every three backbone units (*9*), as shown in Figures 3.1 and 3.2.

The governing factor of the gum's performance is the viscosity that a solution of the gum produces. Guar gum forms viscous, colloidal sols when hydrated in water. Viscosities are usually measured with rotational, shear-type viscometers, such as the Brookfield Syncro-Lectric. These sols exhibit different viscosities at varying shear rates; this is typical of non-Newtonian fluids.

The most highly purified guar gums, food grades, produce viscosities (at 1% solids) of 3000-5000 cps, measured at 25°C with Brookfield viscometer at 20 rpm. Technical and industrial grades may vary from 5000 to less than 500 cps.

An additional factor influencing the performance of the guar gum is its physical form. Since the guar gum retains some of its particulate characteristics when it hydrates, or dissolves, in water, the mesh size of the processed gum affects the appearance and the behavior of the gum solution. Mesh size, to some degree, has a marked influence upon such properties as the rate of viscosity develop-

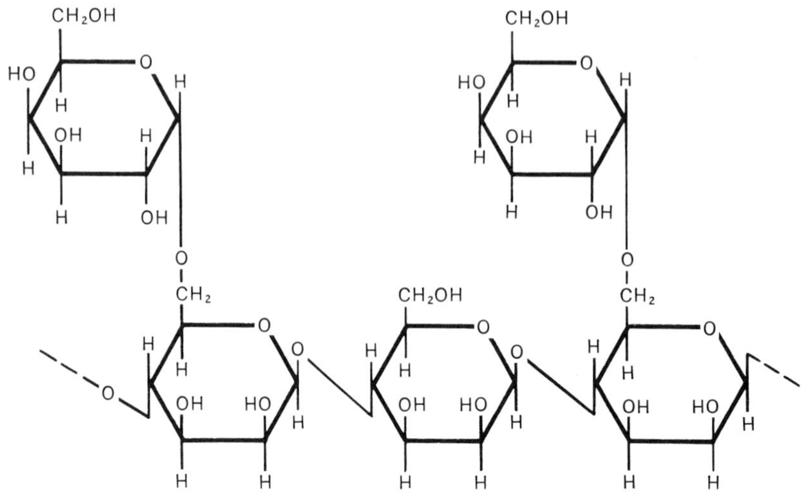

Fig. 3.1. Structural formula proposed for GUARAN, the gum component of the guar bean endosperm, a D-Galacto-D-mannoglycan (ratio ~ 1:1.5).

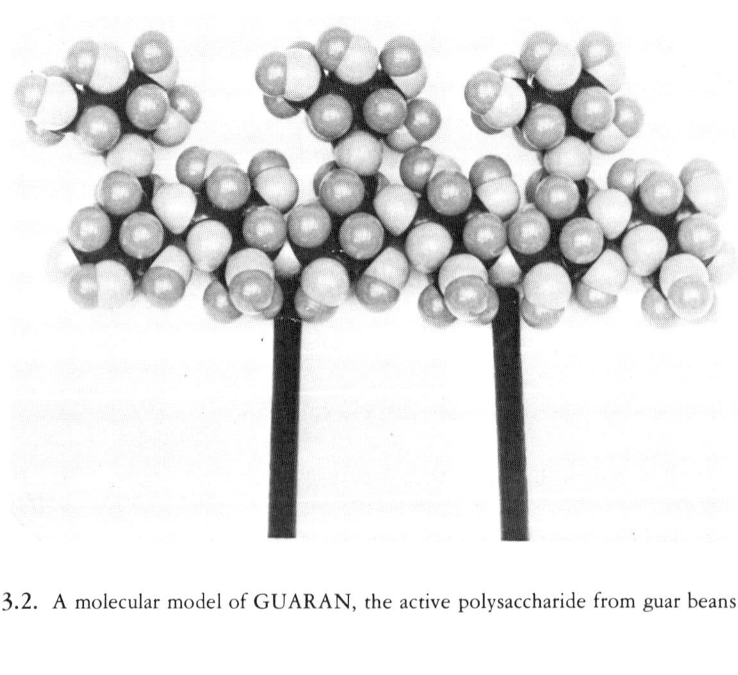

Fig. 3.2. A molecular model of GUARAN, the active polysaccharide from guar beans.

ment, smoothness or coarseness of the solution, and dry flow properties (important for automatic and/or continuous dissolving equipment).

Most grades used in the paper and board manufacturing process are compounded with chemicals which aid in dispersion, control the rate of hydration or achievable viscosity, cause or prevent gelling, and act as preservatives. Some of these properties can be controlled by processing techniques, and thus specific grades of guar gum can be tailor-made for special applications.

Most suppliers listed in this section provide several grades and modifications which are available as standard production items.

Locust Bean Gum

Locust bean gum is derived from the seed of the locust tree (Ceratonia siliqua), a leguminous plant. This plant is a pod-bearing, evergreen, high-staked tree which can be up to 30 feet tall. It has extremely long roots that may penetrate 60 to 90 feet into rocky or stony soil unfit for most other trees. It is native only to the Mediterranean area and has been cultivated for centuries B.C. (*8*). The tree continues to produce pods for several hundred years.

The locust seed is composed of the following (*8*):

Endosperm	35-45%
Germ (Embryo)	25-30%
Seed-Hull	30-35%

The seed must be processed for the removal of the hull and germ by specific milling techniques. The color of the final locust bean gum depends upon the thoroughness with which the seed-hull is removed. The dehulled seeds are then split, and the germ is separated from the endosperm. The cleaned endosperm containing the galactomannan is then ground to desirable mesh size. Depending upon the purification process and degree of refining, four grades of locust bean gum can be obtained: top high-grade, normal high-grade, industrial grade, and technical grade (*8*).

The galactomannan of the locust bean endosperm contains fewer D-galactopyranosyl units per backbone unit than does the one from the guar bean, as shown in Figures 3.3 and 3.4. The galactan to mannan ratios vary from 1:3.0 to 1:6.1 (*10*).

Almost all factors influencing the quality, performance and/or effectiveness of guar gum listed above are shared by the locust bean gum. Due to the larger amount of darker color and a greater tenaciousness of the seed-hull to the locust bean's endosperm, it is more difficult to remove all of the seed-hull, and the final gum may contain a certain amount of dark specks. It may also contain similar quantities of the yellowish germ.

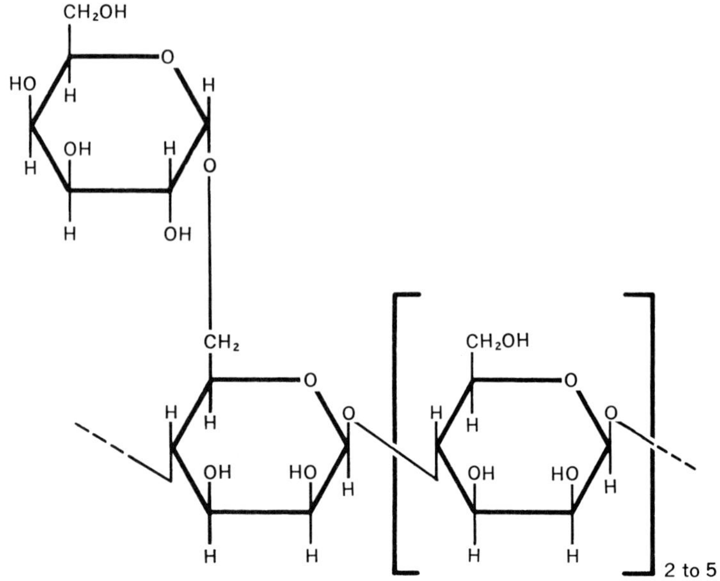

Fig. 3.3. Structural formula proposed for CAROBIN, the gum component of the locust bean endosperm, a D-Galacto-D-mannoglycan (ratio ~ 1:3-6).

Fig. 3.4. A molecular model of CAROBIN, the active polysaccharide from locust beans.

Because of the lower proportion of galactosidic side-branching in its molecular structure, the galactomannan from the locust bean is noticeably less soluble in cold water than is the galactomannan from the guar bean. When the temperature of a dispersion is raised, a momentary decrease in viscosity is followed by a sharp increase. That increase begins at a temperature of about 45°C and continues at rising temperatures until virtually all insoluble particles are dissolved at about 80-95°C. Dispersions heated to about 95°C and then cooled have a much higher final viscosity than dispersions made in cold water.

The most highly purified locust bean gum, food grade, produces a viscosity after heating for a 1% solution of over 5000 cps, measured at 25°C with a Brookfield viscometer at 20 rpm. As with guar gum, the less purified grades of locust bean gum generally yield lower viscosities.

Tamarind Kernel Gum

Tamarind kernel gum is derived from the seed of the tamarind tree (Tamarindus indica), a pod-bearing, very large evergreen that grows rapidly and lives for more than 100 years. It measures up to 80 feet in height and can obtain diameters in excess of five feet. It is one of the most important common trees of the Asian subcontinent and is also found elsewhere in Asia, Africa, and even in Florida. The usefulness of the seed's endosperm as a source of gum was recognized during World War II and has come into commercial use since then. Because there is a great abundance of tamarind seed and the properties of tamarind kernel gum are noticeably lower than those of the former two seed gums, the prices for the various tamarind gum grades are lower.

The tamarind seed is composed of the following (8):

Endosperm (Kernel)	~60-70%
Germ (Embryo)	up to 20%
Seed-Hull (Testae)	20-30%

The tamarind kernels, however, contain up to 7% of an unsaturated oil and up to 22% of a pectin-like proteinaceous material. The presence of the oil (much of it is unsaturated) causes the powder to turn rancid upon prolonged exposure to air. Solvent extraction of the powder yields a virtually oil-free tamarind gum with good shelf life.

As with the previously discussed seeds, the tamarind seed requires processing. However, the preparation of tamarind gum is a comparatively simple operation and consists of dehulling the seeds and grinding the endosperm. As with the locust beans, dehulling is the most difficult step, because the hull is tenaciously held to the endosperm. Through differential grinding, followed by sifting, one can also separate the pectin-like protein from the polysaccharide gum, which is not of a galactomannan type but is a complex molecule consisting of a backbone of D-glucose (3 moles) to which D-xylose pendants are attached (2 moles). To one of the D-xylose pendants one D-galactose unit is attached, as shown in Figure 3.5 (*4,10*). However, the exact structure for this polysaccharide has not yet been verified.

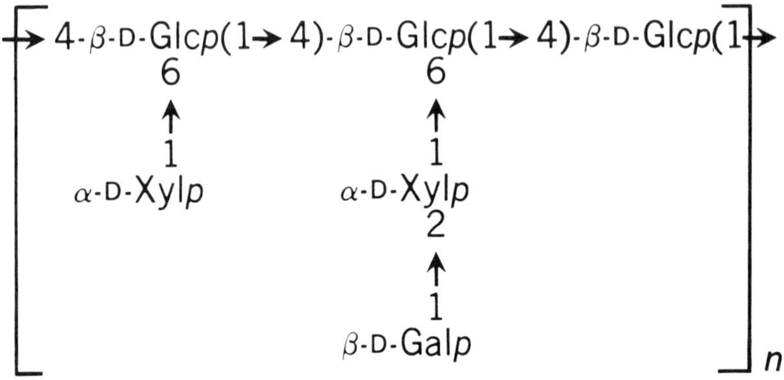

Fig. 3.5. Structural formula proposed for TKP-polysaccharide, the gum component of the tamarind seed endosperm (kernel).

Because of this difference in structure, tamarind gum disperses quickly in cold water, but a uniform solution is obtained only upon heating of the suspension. The maximum viscosity is attained after boiling for 10-20 minutes. A standard industrial grade of tamarind gum produces a viscosity after heating for a 2% solution of about 350-400 cps, measured at 25°C with a Brookfield viscometer at 20 rpm (*11*).

MECHANISMS OF STRENGTH DEVELOPMENT

The dry strength improvement of paper furnishes through the use of polysaccharide gums appears to be derived from the linear nature of the polymer and through hydrogen bonding of the hydroxyl-hydrogen of the polymer with similar functional groups on the surface of the cellulosic fiber. Early investigators (*12*) who researched the reasons for the increase in paper strength when gums are added suggested that mainly the following four factors are involved:

1. Strength of fibers,
2. Strength of fiber-to-fiber bonds,
3. Number of bonds (bonded area), and
4. Distribution of bonds (formation).

An interesting study was conducted in the early 1960's with the aid of 2- and 3-dimensional electron micrography (*13*). The investigators were able to classify the various types of fiber bonds as they relate to pulp fibers, refining, and drying conditions. They postulated that gum additives, provided they are retained by the fiber, increase the strength of the paper by positively influencing one or more of these four factors. It was proposed that gums improved the paper strength by influencing each of the individual factors by the following percentages:

Changed strength of fibers	0%
Attributable to fiber-to-fiber bonding	60%
Attributable to number of bonds (bonded area)	15%
Attributable to formation	25%

At the present time it has not yet been determined whether the greater bonding strength results from greater intrinsic bond strength between the gum molecules and cellulosic fibers, greater bond flexibility, or a greater number of bonds per unit area of bonds (*14*).

More recent findings revealed that shortening of the polymeric chain through hydrolysis or depolymerization or extensive etherification of the hydroxyl groups significantly lowers a gum's dry strength improving potential (*15*). It was further found that the proportion of dry strength improvement attributable to deflocculation of the suspended fibers and greater interfiber bonding strength after sheet formation depends greatly on stock (furnish) variables such as consistency, refining, and fiber length.

Studies have shown that the various gums work quite well and effectively on a given paper furnish in the laboratory, but in the mill the performance may be disappointingly poor. The explanation in a given instance depends, naturally, upon a variety of furnish and machine conditions. In some cases there is no explanation. However, in a majority of cases, a close examination of the entire mill system has resulted in surpassed or expected mill performance (*16,17*).

Further research was carried out to determine the specific sorption of the gum onto the pulp fiber. This work involved the sorption of radio-carbon-14 tagged gum onto a moderately refined softwood sulfite pulp and the preparation of handsheets therefrom. The fines in the white water and the white water itself were also analyzed for the presence of radio-carbon-14. That study showed that at low gum dosages (0.1 to 0.5% based on the fiber) the total sorption of gum onto the fiber is high. However, it was also found that at 0.1% gum (2 pounds per ton of fiber) the fines sorbed 24% of the gum in the system. This means that the surfaces of the longer fibers sorbed only 0.07% gum (1.4 pounds per ton of fiber), and the fines sorbed the remaining 0.03% (which amounted to 40 pounds per ton of fines). Thus, the small amount of fines in the system does end up with a more thoroughly gum coated surface. Similarly, it was found that at higher dosage levels the fines sorbed a significant fraction of the gum. In general, both rate and amount of sorption of gums are direct functions of the specific surface and contact time.

It was further determined that clay and other high surface area constituents of the papermaking furnish also effectively compete with the long fibers for the gum in the system. Therefore, it becomes quite plausible that furnishes which contain large quantities of fines, fillers, or secondary or recycled fibers generally respond poorly if gums are added to the whole furnish.

However, if the gum is added to the furnish before clay or recovered or recycled fibers are introduced, the effectiveness of the gum is greatly improved. This practice permits the longer fibers to sorb a maximal amount of gum onto its surfaces, where it does the most for improving the dry strength of the finished sheet (*18*).

BENEFITS AND ADVANTAGES

Dry strength additives are used primarily for improving bonding strength between papermaking fibers. However, this benefit may be utilized in various ways, among which may be the following:

1. Improvement of bursting and tensile strengths, folding endurance, as well as surface characteristics, by adding a small percentage of gum directly to the refined pulp;
2. Utilization of a higher amount of short-fibered pulp or secondary (recycled) fibers and pigment fillers at initial sheet strength;
3. Reduction of energy consumption in the refining or beating process;
4. Maintenance of tearing strength through reduced refining;
5. Increase in mill capacity by reducing pulp refining requirements;
6. Boost in paper production through acceleration of the machine, permitted by less refining, and enabling improved water removal on the wire (drainage);
7. Improvement of fiber distribution or sheet formation through the use of gum additives; and
8. Attainment of combinations of properties otherwise difficult to obtain in a given furnish. For example, high strength and high porosity in paper is a combination unobtainable without the use of gum additives.

It was shown by several investigators that the more efficient gum additives are physically adsorbed onto the surfaces of the cellulose fibers, and it is quite clear that the layer of the highly solvated polysaccharide on the fiber surface improves bonding by enhancing molecular contact of the surfaces and by forming bridges between the more distant surfaces. It was also postulated that Campbell-forces during drying might move the well dispersed gum molecules toward fiber cross-over areas (*19*).

The above findings are demonstrated in the following scanning electron micrographs (*20*). Figure 3.6 shows the surface of two adjacent bleached western softwood fibers. The surface shows in great detail the microfibril structure and the complete lack of any interconnections between the two crossed fibers which could initiate interfiber bonding. Figure 3.7 depicts the same bleached western softwood pulp at a consistency of 0.5% pulp to which 0.02% (0.4 pounds/ton) guar gum was added. The suspension was allowed to air dry on a slide, and a specimen for the scanning electron microscope was prepared from it. The micrograph shows the pulp fiber to be totally enveloped with a thin layer or coating of guar, and, in addition, the guar was filmed across to an adjacent fiber, forming a solid bridge or web.

Figure 3.8 presents the same fiber at a consistency of 0.5% pulp to which 0.02% (0.4 pounds per ton) locust bean gum was added. The slide was prepared as above. This micrograph shows the pulp fiber to be coated even more thoroughly than when guar is used. The microfibrils on the fiber's surface are totally covered with a thick layer of locust bean gum coating. Apparently due to a more dense adsorption onto the fiber, less webbing or bridging occurred (*20*).

USE OF GUMS IN THE PAPERMAKING PROCESSES

Generally, the paper furnish at an average consistency of about 0.5% (0.25-1.25) is pumped by a fan pump to a headbox. A solution of about 1% gum at a dosage level of up to ~1% of gum based on the pulp is usually added to the fan pump, where thorough mixing occurs. Recycled white water from the wire is mixed at the fan pump with the fresh thick stock from the stock meter. This recycled water contains about 0.2% (0.02-0.65%) fiber-fines which have drained through the wire. Rough estimates and years of operating observations have shown that the volume rate of the recycled white water is about 5-20 times the volume of fresh stock entering the system.

As a general assumption, the concentration of gum can build up in the white water to about 5 times the dosage rate or to about 2.5% based on the pulp (0.0125% based on the suspension). This means that the fines in the white water have been exposed to such a high concentration of gum for approximately 2 to 10 minutes. From the fan pump, the furnish/white water mixture goes to the headbox and out onto the wire. The time of contact of the gum with the freshly added fibers between the fan pump and the wire is about 5-15 seconds.

Hence, these fresh fibers are only partially surrounded or saturated with gum, whereas the fines in the white water are near total saturation because of the long contact time. This means that the kinetics or rate of adsorption are very

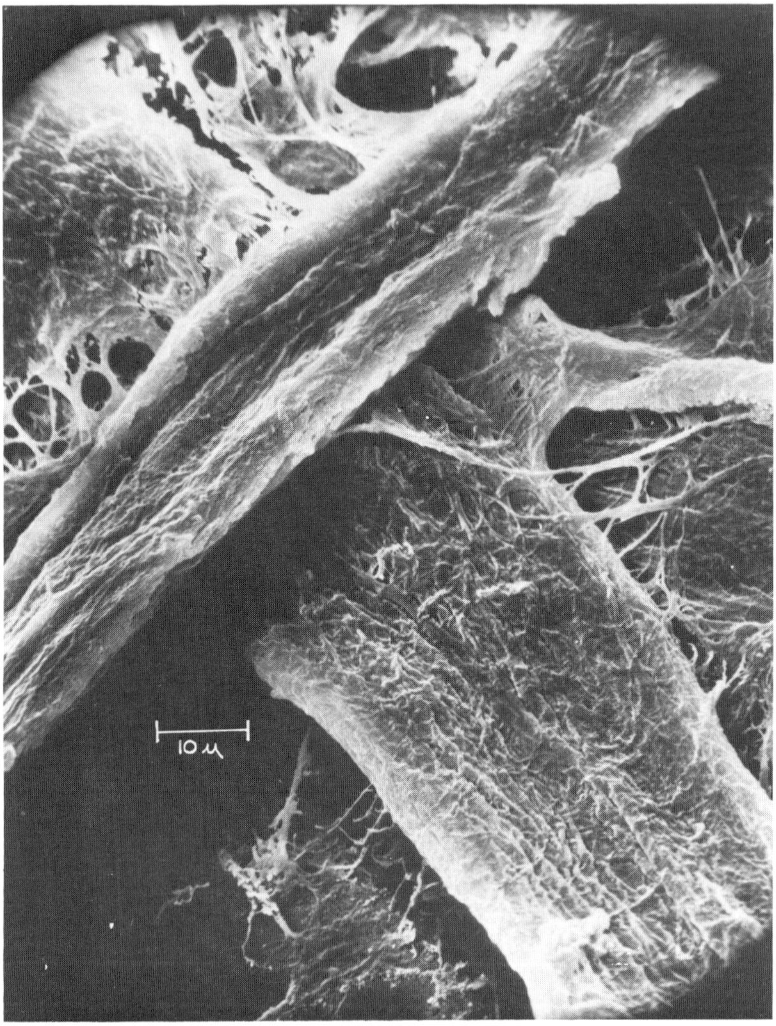

Fig. 3.6. Scanning electron micrograph of air-dried bleached western alpha pulp without additives. Large fibers and fine fibrils are shown in great detail with no fiber-to-fiber bonding discernible (magnification 2500×).

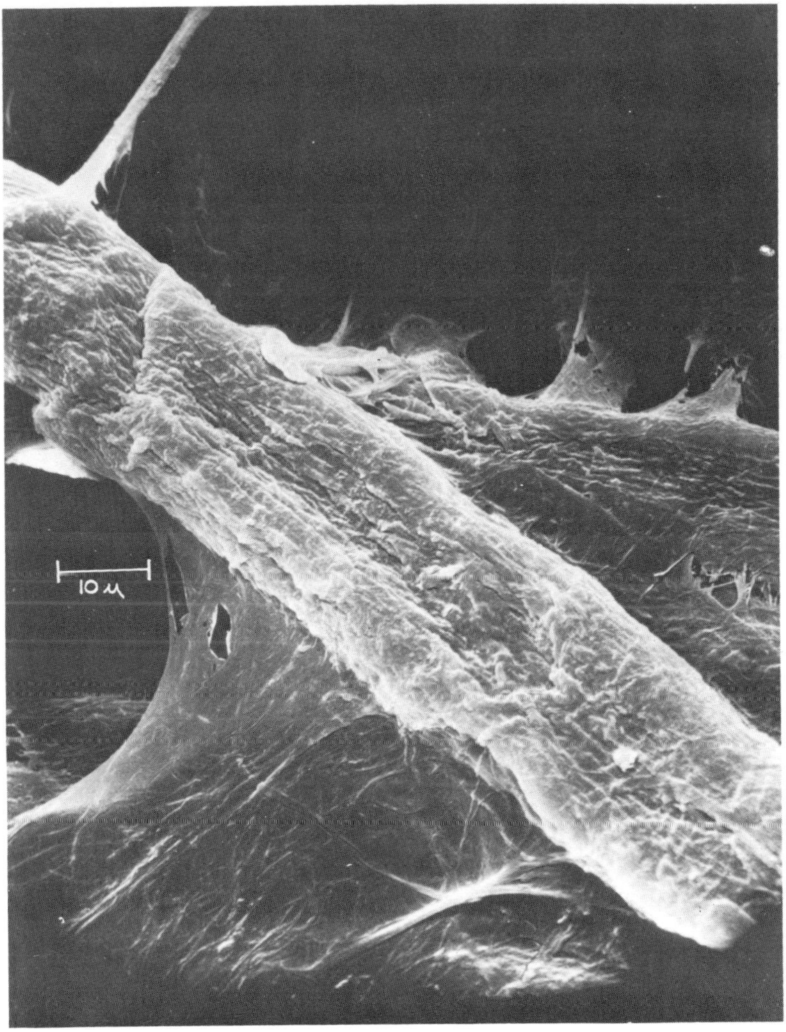

Fig. 3.7. Scanning electron micrograph of air-dried suspension of bleached western alpha pulp tc which 0.02% guar gum was added. The coating on the fibers covering the fine fibrils and a film of guar gum connecting two fibers at a cross-over point are clearly shown (magnification 2500×).

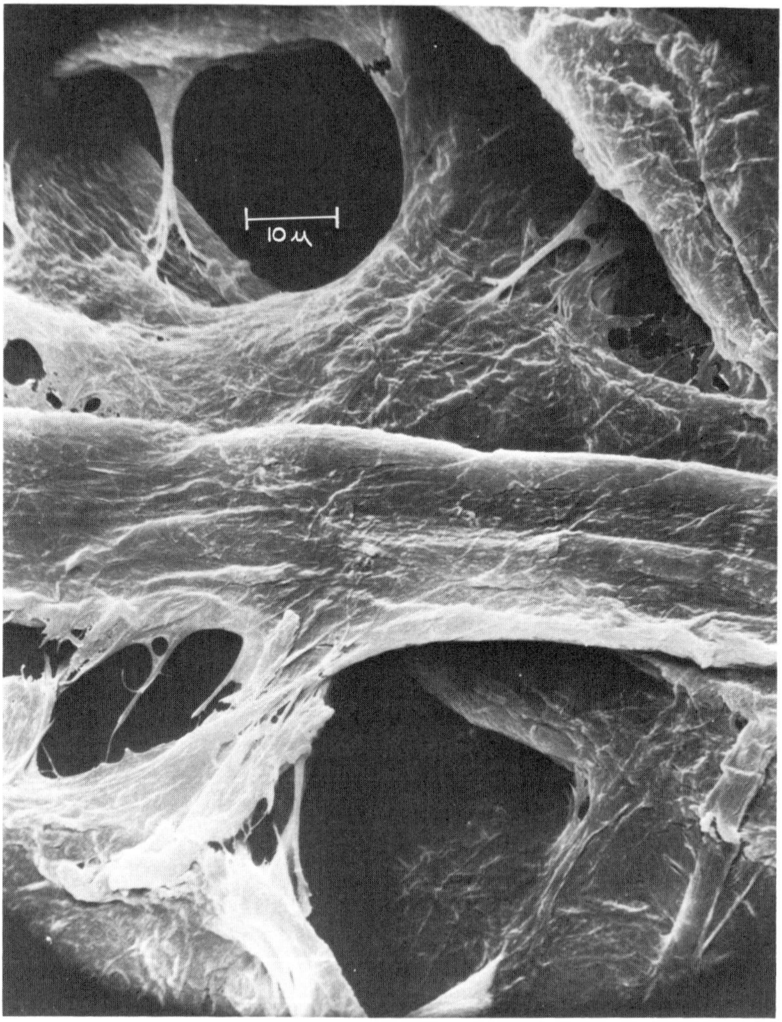

Fig. 3.8. Scanning electron micrograph of air-dried suspension of bleached western alpha pulp to which 0.02% locust gum was added. It shows a much thicker coating on and less webbing between the fibers, an indication that more polymer is being adsorbed onto the fiber surfaces (magnification 2500×).

important factors to be considered when using gums. It should be realized that the numbers given above are only approximations, because different machines differ widely in their processing and recycling ratios, and buildup of gum in the system is usually not one of the measured or controlled quantities.

Depending on the paper machine system and the stock preparation equipment supplying the respective machines, an optimum location for the addition of gums must be found through a series of preliminary mill trials.

Figure 3.9 shows a schematic presentation of fourdrinier and cylinder type paper machine systems. It further shows the many possible locations where gums can be introduced into the paper furnish for internal addition or applied to the formed paper to improve surface characteristics (*21*). Invariably, the question regarding the best points of addition for gums arises. As noted in the preceding paragraph, only a series of mill trials can point to and help determine the optimum location for a specific machine or furnish. Some mills have found that different furnishes processed on the same paper or board machine require different gum addition points.

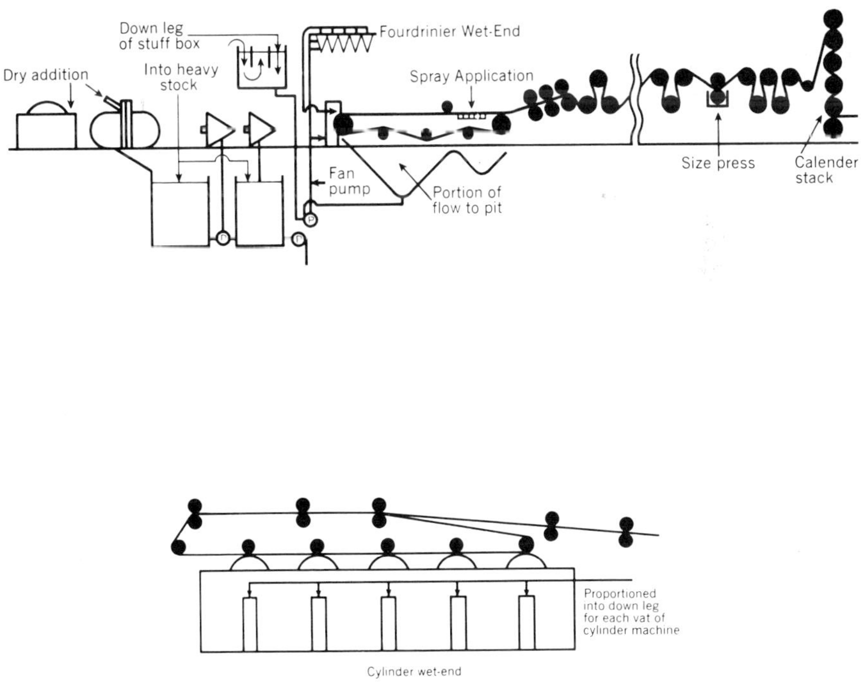

Fig. 3.9. Schematic diagram showing various gum addition points in stock preparation areas, at fourdrinier and cylinder-machine wet ends and for surface applications to the finished sheet.

However, generally speaking, gum solutions may be added to any of the several locations noted in Figure 3.9. Gum solutions may be added to either the dilute stock at the fan pump or to the return white water line in fourdrinier systems, or both. On multi-cylinder board machines, a prevalent method is to add the gum solution to each of the stock lines that supply the vats and at the point of desired use. It could be added to the headbox, before each vat, to the stock-dividing system, or to the stock chests. One of the most important aspects of gum addition to any stock system is that the dosage rate must be constant and in exact proportional amounts to the stock flow.

Regarding dual headbox fourdrinier board machines, one has the option of adding the gum solution to both of the stocks in amounts found to be most beneficial for each stock or of adding all of the gum requirement for the whole furnish to either system. In some cases it was found that adding about half of the total required gum only to the secondary furnish obviated the need to add any gum to the primary furnish.

Various types of gums can also be introduced to a paper furnish in their dry form. However, this can be done only with specially formulated types of gums and in stock preparation systems where equipment, stock conditions, retention times, and certain other necessary prerequisites assure proper conditions for the complete dispersion, dissolution, and sorption of the added gum. Such additions are usually made to the make-up water in batch-type beater systems or waste paper repulping equipment. This procedure is especially recommended when the repulping process utilizes water at temperatures in excess of 55°C (130°F) and no gum cooking or solution make-up systems are available. It must be noted here that the use of properly dissolved (cooked) gums is always preferred and recommended for best effectiveness, efficiency, and economy (*22*).

Another technique for internal addition of gums is spraying a solution of it onto the partially consolidated wet web on the fourdrinier wire. This is done by positioning a sheet-wide header with uniformly spaced spray nozzles near or in the area of the vacuum boxes and by applying a well prepared and thoroughly filtered gum solution in the appropriate dosages to achieve the desired effect. This type of application is especially well suited for light weight papers such as offset newsprint. Other grades that may benefit from such an application of gums are in the linerboard category. A spray application of gum solution to the primary sheet just ahead of the secondary headbox promotes the interfiber bonding of the fibers deposited from the secondary headbox to the fibers already consolidated on the wire.

Finally, gums are used as surface modifying agents in both paper and board manufacturing. Application of gum solutions usually takes place at a size press and, in some cases, at calender stack boxes. Gum may be used alone for the sealing-in of short surface fibers where no actual coating effect is needed or desired. Such applications are also indicated when the curl properties of a one-side coating base sheet require improvement.

Detailed technical brochures are available from the various suppliers of gums listed in TAPPI CA Report No. 60. These brochures deal with almost every aspect of gum use for the improvement of dry strength and related properties in paper and board manufacturing processes.

HANDLING, PREPARATION, AND FEEDING OF GUMS

Handling and Storage

Specific care should be exercised in handling gum products. Usually, the manufacturer's recommendations call for the storage of the product, shipped in multiwall bags and stacked on pallets, in cool, dry places.

There are demonstrated reasons for such requirements. First, these gums swell readily in water and produce an extremely slippery mucilage. Secondly, spills of dry gum on damp or wet floors quickly produce a slippery condition which is to be avoided under all circumstances. An excellent practice is to store the main supply of gum away from the make-up equipment in dry warehousing areas.

Similarly, as with synthetic polymers, dry gum spills should be promptly removed or vacuumed, not washed off, in dry floor areas. Fine gum particles can stick to one's shoes and turn into a slippery coating upon entering wet floor areas. However, wet floors should be immediately hosed down with water to completely remove all slipperiness.

Preparation

Gum solutions can be prepared in a number of ways, depending on equipment available at a particular paper mill.

One of the simplest solution preparation methods is a batch system, as shown in Figure 3.10. Such a system usually consists of a tank equipped with an agitator, water supply line, and a sifting screen with a small platform attached to the rim of the tank. Whenever a specific type of gum needs to be cooked, a steam line with exit-spargers should be provided near the bottom of the tank.

When equipment limitations prevent adequate agitation in the mixing tank, an eductor should be used instead of a sifting screen, as shown in Figure 3.11. The eductor uses high velocity water to draw the gum from a supply funnel and causes quick and thorough dispersion of the gum in the mixing tank (*23,24*). Inadequately dispersed gum suspensions, however, may result in incomplete dissolution of the gum, which could cause imperfections (fish-eyes) in the finished paper. Reduced efficiency of such gum solutions may lead to increased consumption, further compounding the problems.

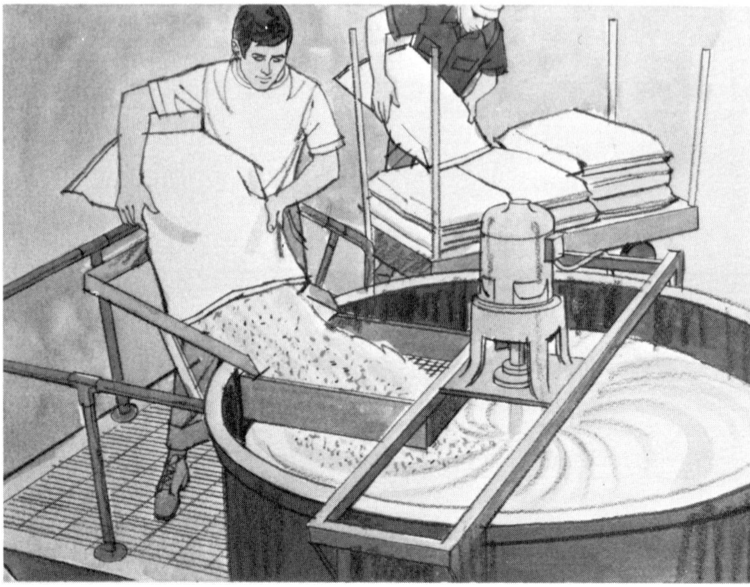

Fig. 3.10. Manually operated batch tank system, the simplest gum make-up system.

Some papermaking processes consume extremely large volumes of gum solutions. This is primarily due to three fundamental factors: high machine production output, use of large percentages of recycled or short fibers, and a combination of the above. When it is necessary to provide very large amounts of gum solutions, a continuous make-up system is indicated. There are several systems used for continuous gum solution preparation.

One such system utilizes a hopper attached to a continuous dry feeder, which discharges the gum to a pre-slurry or make-up tank, as shown in Figure 3.12. Both gum and water enter that tank continuously in appropriate proportions. The pre-slurried gum then overflows into another, larger tank, which is also equipped with an agitator. In addition, a steam supply may be provided where expeditious and complete dissolution of the gum is desired.

For gums that require cooking, several types of continuous cooking systems which achieve complete dissolution of the dispersed particles are available (*23,25*). The simplest systems consist of a continuous solution preparation unit as described above, to which one or more steam cooking tanks are added, as shown in Figure 3.13. The slurried gum suspension from the make-up tank flows into the first chamber of a series of compartments that are heated with live steam. From there, the gum solution cascades into the succeeding chambers

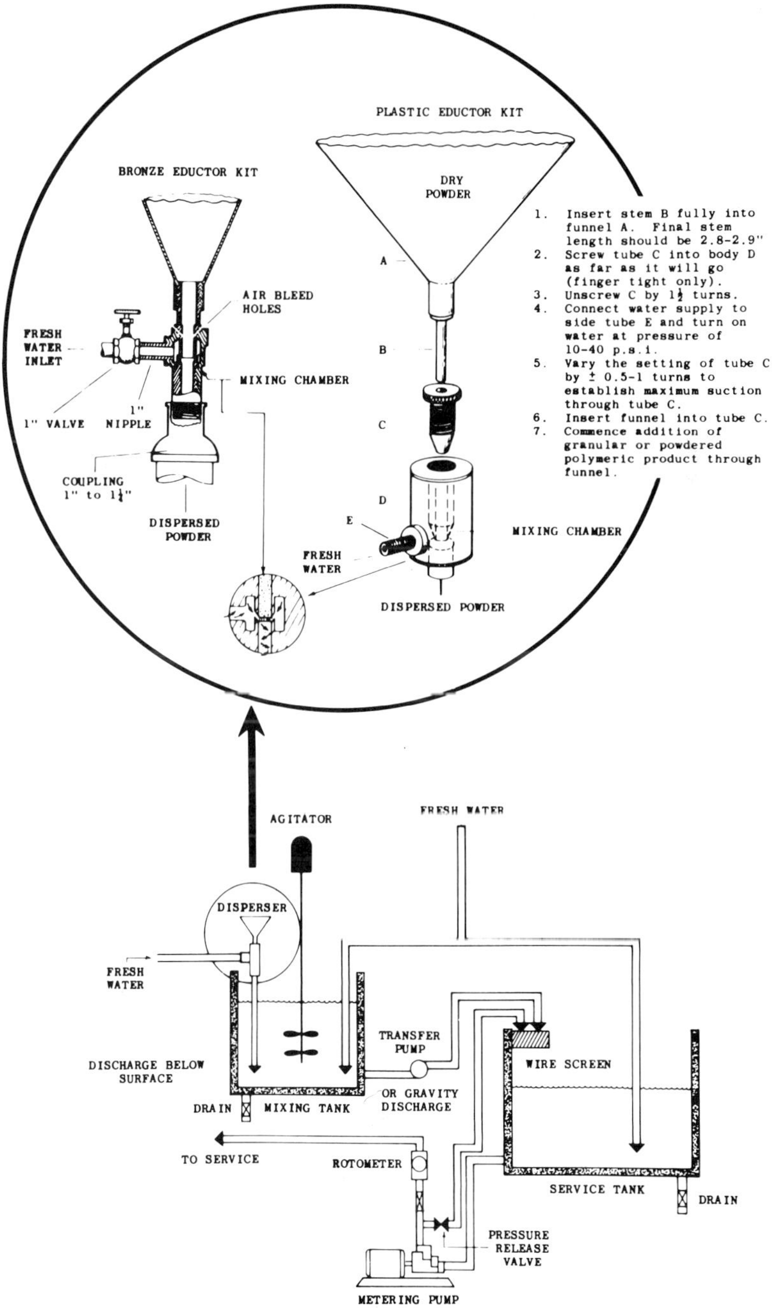

Fig. 3.11. Typical solution make-up and metering system for granular or powdered gum products, utilizing a disperser (eductor) for thorough wetting and dispersing.

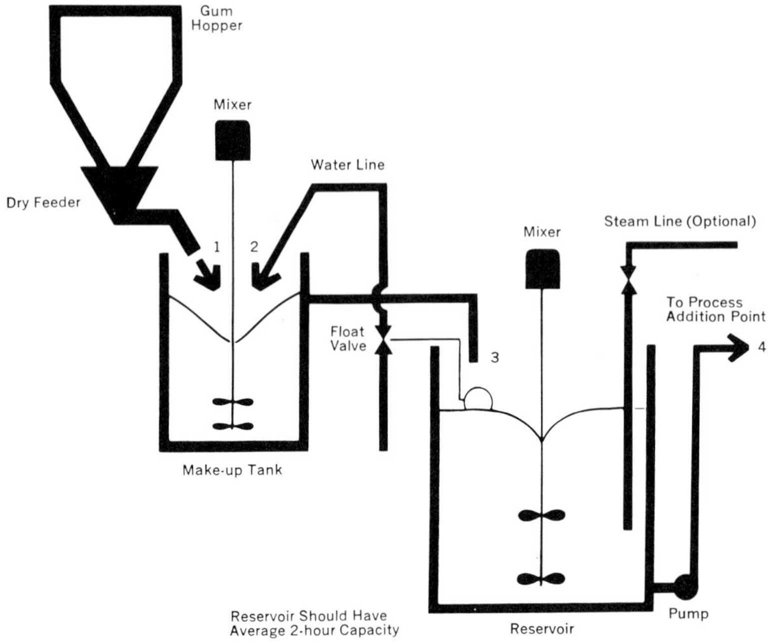

Fig. 3.12. Simplified schematic of a continuous gum solution make-up system for cold solution preparation. Dry gum discharges into vortex at 1, and water flows in at 2. The pre-slurried gum dispersion overflows into service tank at 3. The gum solution is metered to process at 4.

where it is progressively heated until it reaches about 85°C (~190°F). Only a relatively unsophisticated control mechanism is needed if it is desired to convert such a cooking system to automatic operation.

Gums are usually slurried up and cooked at 0.5 to 1.0% solids and used without further dilution. Actual concentrations may depend upon such factors as the gum's viscosity potential, machine requirements, and the capacity and sophistication of the make-up system. When using high grade, high viscosity gums, especially for fine paper grades, care must be exercised to maintain a concentration near 0.5%, to keep the pre-slurry water temperature below 30°C (85°F), and to maintain the cooked gum solution at 75°C (170°F) or higher (*21*).

In recent years, fully automated jet cooking systems have come into use for the preparation of gum solutions. However, it has been found that very complex and carefully designed control mechanisms are needed to make their operation trouble-free (*24,26*).

When the necessity for adding dry gum to a papermaking system arises, usually at a pulper or beater, it is preferable to sift or educt the required quantity as quickly as possible into the make-up water under most vigorous agitation.

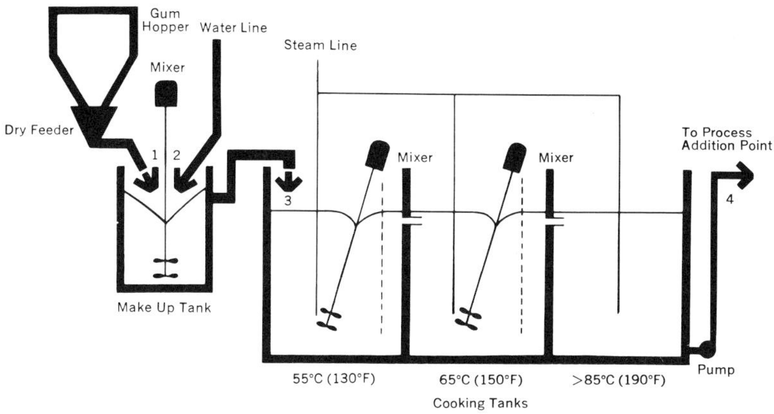

Fig. 3.13. Simplified schematic of a three-compartment continuous gum cooker. Dry gum discharges into vortex at 1, and cold water flows in at 2. The pre-slurried gum dispersion overflows into the first cooking compartment at 3. From there, the gum solution overflows into the succeeding compartments, where it is progressively heated to the final temperature of over 85°C. From the final chamber, the gum solution is pumped to a service tank or metered to the process at 4.

The waste paper or pre-slurried pulp should not be added until full dispersion of the gum is achieved. This practice prevents the formation of lumps which could subsequently show up as imperfections (fish-eyes) in the paper. However, when dry gum is to be added to a slush-stock, greater care during addition should be exercised, especially when consistencies are substantially above 3.5 percent.

The utilization of dry gum addition is not recommended in a paper stock sys tem where the total retention time from point of addition to the headbox is less than 90 minutes. However, when the stock preparation process is operated at temperatures above 55°C (130°F), a total retention time of not less than 45 minutes is desired.

Feeding and Metering Gum Solutions

The most reliable gum solution feeding systems employ adjustable positive displacement pumps. Usually, when several paper machines are to be supplied with gum solution, a central supply tank feeds the solution to a main header. From there, individual metering pumps take the gum to the respective machine systems.

When selecting metering pumps, it is essential to consider such factors as gum solution viscosity, pipe diameter and number of fittings in the line, flow resistance, head differential between the central supply tank and the discharge point at the machine, total distance from pump to machine, and discharge of solution into open stock boxes or into closed stock lines that are under pressure (*24*).

Some installations employ centrifugal pumps that feed the gum solution under pressure to a looped supply line with the excess returning to the central supply tank. Individual taps in the looped gum supply line allow for relatively accurate metering of the solution to each specific addition point. Such supply loops, however, should not be used to supply gum solutions to size press or calender box solution make-up systems that employ a batch system. Such make-up systems should be supplied directly from the central tank to avoid sudden pressure drops (causing flow reduction to the machines) in the looped line.

Of the various types of flow meters that are suitable for metering gum solutions to a machine, glass tube and float flow meters are in most general use. Magnetic flow meters with flow indication dials are also used, but because of their much higher cost they have not found as wide an acceptance. Paper machines with fully automated and computerized wet end systems can interlock the gum addition system with the automatic basis weight and machine speed control systems. This ensures a continuous supply of gum solution to the stock system at the exact required or desired proportions.

Gum solutions should not be gravity fed to a stock system. Wide variations of solution temperature cause changes in viscosity and may result in nonuniform flow to the machine.

SPECIFIC USE APPLICATIONS (GENERALIZED CASE HISTORIES)

Based on the benefits that a papermaker can achieve — improved bonding, strength (tensile, burst, scuff resistance, crush, folding endurance) with maximum machine output, noticeable power savings — gums have established themselves as an important additive in the paper and board industry.

In the manufacture of test liner, corrugating medium (both NSSC and bogus), jute liner, bleached kraft board, coated cylinder board, machine glazed bag and wrapping papers, multiwall bag papers, newsprint and offset newsprint, and many other specialized grades of paper or board, gums are now accepted as a valuable additive which enables paper and board mills to meet highly competitive market conditions. This is becoming increasingly necessary with the greatly expanded utilization of waste paper and recycled fiber in paper or board furnishes where hitherto such uses were foreign (*27*).

Kraft Linerboard

In total volume, kraft linerboard mills are reported to be the largest users of the various types of gums. The particular geographic areas of use include mills in the South and Southeast, as well as the Pacific Northwest, of the United States. While gums are generally used to improve bonding and strength properties, the specific benefit to these mills is the assurance of maximum possible bursting strength at maximum machine operating speeds. With the ever-increasing demand for strong containers, it is mandatory to obtain maximum possible strengths at minimum allowable basis weights (*15,27,28*).

In this particular grade, dosage rates of gum range from 0.05 to 0.3% (1 to 6 pounds per ton), but southern kraft mills rarely use more than 0.2% (4 pounds per ton). The general preference in these mills is gum derived from guar beans.

The fan pump is the preferred addition point in these mills, and the gum is added as a cooked, dilute solution of about 0.25 to 0.75% solids.

On machines having secondary headboxes, the gum solution is preferably added to the primary stock whenever a great difference exists in the freenesses between the primary and secondary furnishes. The gum slows the primary stock on the wire and therefore allows for more thorough penetration of the secondary stock, thus increasing ply bond. This becomes very important where difficulty in meeting scuff resistance is related to secondary peel (*28*).

The use of gums in the primary stock also aids in sheet formation, thus presenting a more uniform rate for the water of the secondary stock to filter through, a point which papermakers consider to be very important (*28*).

The resultant benefits usually include improved burst and machine speed increases of 50-100 feet per minute. Strength improvement is especially noticeable in furnishes that contain larger proportions of hardwood kraft pulp from gum and oak trees (*15,27*).

Bleached Kraft Board

Bleached kraft board is usually produced as a solid sheet; that is, the furnish throughout the sheet is the same. These sheets may require gums in order to meet and maintain strength specifications while keeping up maximum production rates. With bleached southern kraft pulps, changes in operating and sheet characteristics are generally in direct proportion to the amount of gum added; drainage rate drops and wet end vacuum (couch) increases, indicating a more uniform and tighter formed sheet. Normally, when this occurs, refining can be reduced to restore the drainage rate. Occasionally this can be accompanied with an increase in machine speed (*22,28,30*).

One grade that particularly benefits from the addition of gum to its furnish is cup-stock made on fourdrinier machines. Here the advantage of having used gum in the process emerges in the converting operation. Whenever it was necessary to produce cups with rolled or lipped edges, it was found that the stock had appreciably improved ply bonding and bending properties (*28*).

Corrugating Medium (NSSC and Bogus)

The use of gums in the production of corrugating medium affords the best possible means of improving the flat crush or "concora" test (*11,15,31,32*).

Mills that use gums for this purpose are widely scattered and are located throughout the United States, Canada, Europe, Japan, Australia, and New Zealand.

In the past, gum from locust beans was more widely used, but lately gums from guar or tamarind seed have found acceptance (*11,15,30,31*). It was felt that locust bean gum would be more suitable because of its higher sorptivity onto the secondary fiber at the usually prevailing higher pH conditions (about 8 to greater than 9), but later it was found that gums from guar beans provide more increase in strength per pound of gum adsorbed. Perhaps this may be due to the larger amount of galactosidic groups which are present in the guar molecule and provide additional bonding sites.

Dosage levels of gum for corrugating medium usually range from 0.3 to 0.75% (6 to 15 pounds per ton), but 0.4% (8 pounds per ton) is generally considered sufficient for furnishes not containing excess proportions of waste paper or recycled fibers (*15*).

In some mills where waste paper (board) or short fibered recycled stock comprise the major proportion of the furnish, gum dosages in the range of 0.6 to 1.2% (12 to 24 pounds per ton) may be encountered. In these applications, the gum can be introduced in two stages. One point of addition is usually at the repulper (with the waste paper board), and the second one is somewhere closer to the wet end of the machine, preferably to the thick stock at the stock meter prior to the fan pump.

Machine Glazed Bag and Wrapping Paper

Much of the paper in this category is produced by smaller paper mills, using dried market pulp or almost exclusively secondary fiber. There are not too many mills fitting into this category in the United States; however, they are numerous in Japan, England, and the European continent.

Since these furnishes require more extensive mechanical refining to achieve adequate and competitive tensile and burst strength, more water is held by such highly refined fibers. This is an undesirable property when a uniform glaze finish is required. For those grades, dry gums are customarily added to the stock preparation prior to refining, usually at the pulper. This practice allows a lesser degree of refining, while maintaining uniform fiber distribution (formation) which results in a wet-spot free glazed finish (*33*).

The customary dosage rates of gums for these grades range from 0.4 to 1.0% (8 to 20 pounds per ton) (*15*).

Offset Printing Papers

The use of gums for these particular paper grades is worldwide. However, the most effective uses were developed in Canadian and Scandinavian newsprint mills. The specific grades that were developed with the help of gums are improved groundwood furnished web offset printing papers. They were developed as economical, yet high quality, multi-color printing papers for uses where permanence is not essential.

Improvement in the quality of groundwood and selection of the proper chemical pulps still fell short of solving the major problem for most high percentage groundwood papers—namely, linting, the shedding of only partially bonded, short, surface fibers (*15,29,30*). These loosened fibers foul the printing blankets and require frequent shutdowns for cleaning purposes.

The use of gums in these grades has been found to be very effective in reducing linting problems (*30*). Papers made with gum reduced the number of cleanups by a factor of two to three times, as compared to papers made without gums.

Usual dosage rates of gum for linting reduction, improved surface fiber bonding and favorable deflocculation, are from 0.1 to 0.3% (2 to 6 pounds per ton). Most mills prefer guar gum for this application because of its higher effectiveness at lower dosages (*15*).

Newer Uses of Derivatized Gums

In the past few years, cationically derivatized gums have begun to show up in various paper and board mills. The derivatization of gum imparts to the basic nonionic polysaccharide a cationic functionality.

The most effective cationic gums are of the quaternary ammonium chloride variety (*34,35*). In addition to the basic hydrogen bonding properties that are inherent in the gum, the cationic functionality helps to reduce the usually highly negative zeta-potential of the paper furnish, especially of the fines and filler particles in the white water.

This change of zeta-potential leads to a more thorough agglomeration of the fines in the system by forming more cohesive flocs. These, in turn, are trapped by the longer fibers, thus filling the voids among the larger fibers with additional material that helps in the interfiber bonding of the wet web, which leads to dry strength improvement.

Additional benefits arise from the use of such derivatives by reducing the load of fines in the white water and improving the drainability of the consolidated wet web on the wire. In several cases, machine speed-ups to the limit of the drive mechanism under the most adverse stock preparation conditions were achieved when producing 42 and 60 pound linerboard, while maintaining all strength requirements.

Due to their high retentivity, add-on levels for cationic gums range from 0.05 to 0.25% (1 to 5 pounds per ton) (*34*). About 0.14% (2.75 pounds per ton), added to the thick stock prior to the fan pump, has been found to be effective for burst improvement, reduction of solids in the white water, retention of the newer neutral sizes, and wet web drainage improvement.

Literature Cited

1. B. W. Rowland, *South. Pulp Pap. J.*, 8(7): 20-24 (1945).
2. J. A. Esser, *Chemurgic Digest*, 15(12): 4-7 (1956).
3. J. A. Esser, *Chemurgic Digest*, 17(4): 9-12 (1958).
4. R. L. Whistler and J. N. BeMiller, "Industrial Gums," 2nd ed., Academic Press, New York, 1973.
5. J. W. Swanson, "Chemical Additives: New Dimensions for Papermaking," in "Chemical Additives—Application, Theory and Control," Western Michigan University, Kalamazoo, Jan. 1970, p. 130.
6. W. F. Reynolds, "Commercially Available Chemical Agents for Paper and Board Manufacture," CA Report No. 60, TAPPI, Atlanta, 1976.
7. H. Goldfrank, *Chemurgic Digest*, 16(2/3): 10-11 (1957).
8. R. L. Whistler and J. N. BeMiller, "Industrial Gums," Academic Press, New York, 1959.
9. H. S. Dugal and J. W. Swanson, *Tappi*, 55(9): 1362-67 (1972).
10. F. Smith and R. Montgomery, "The Chemistry of Plant Gums and Mucilages," American Chemical Society—Monograph Series 141, Reinhold, New York, 1959.
11. "Keygum Mark IX," Product Information Bulletin, American Key Products, Inc.
12. H. J. Leech, *Tappi*, 37(8): 343-49 (1954).
13. G. Jayme and G. Hunger, "Electron Microscope 2- and 3-Dimensional Classification of Fiber Bonding," in "Formation and Structure of Paper," ed. by Francis Bolam, Technical Section, The British Paper and Board Industry Federation, London, 1962, pp. 135-70.
14. J. W. Swanson, *Tappi*, 44(1): 142A-181A (1961).
15. W. E. Skagerberg, "Galactomannan Gums and Their Derivatives in Papermaking," 22nd Plastics Paper Conference Preprints, TAPPI, New York, 1967.
16. J. W. Swanson, *Tappi*, 43(3): 176A-180A (1960).
17. A. H. Nissan, "General Principles of Adhesion, with Particular Reference to the Hydrogen Bond," in "Formation and Structure of Paper," ed. by Francis Bolam, The British Paper and Board Industry Federation, Technical Section, London, 1962, pp. 119-134.
18. J. W. Swanson, *Tappi*, 33(9): 451-62 (1950).
19. H. C. Schwalbe, "Effects of Sizing, Adhesives and Fillers on the Formation and Consolidation of Paper Webs," in "Consolidation of the Paper Web," ed. by Francis Bolam, The British Paper and Board Industry Federation, Technical Section, London, 1966, p. 709.
20. C. J. Geankoplis, Private communication, Preliminary Study of Guar and Locust Bean Gum in Papermaking, General Mills Chemicals, Inc. (now Henkel Corp., Minneapolis, MN), 1969.
21. A. J. Winchester, *Pap. Trade J.*, 149(51): 38-40 (1965).
22. "Guartec CAP," Chemicals for the Paper Industry Bulletin, General Mills Chemicals, Inc. (now Henkel Corp., Minneapolis, MN), (A26895).
23. "Guartec Dispersion and Sol Preparation," Technical Information G-89, General Mills Chemicals, Inc. (now Henkel Corp., Minneapolis, MN), July 1962.
24. "Solution Make-Up Procedures," Technical Information Bulletin TI-05, General Mills Chemicals, Inc. (now Henkel Corp., Minneapolis, MN), July 1971.
25. "Continuous Sol Preparation System," Technical Information G-54a, General Mills Chemicals, Inc. (now Henkel Corp., Minneapolis, MN), Jan. 1964.
26. E. J. Loftesness, *Pulp Pap.*, 43(12): 135-37 (1969).
27. "Guartec in Linerboard," Technical Information G-102a, General Mills Chemicals, Inc. (now Henkel Corp., Minneapolis, MN), Sept. 1965.

28. D. V. Lent, *Paperboard Packag.*, 47(1): 135-36 (1962).
29. "Offset Print Quality," Technical News, TN-03, General Mills Chemicals, Inc. (now Henkel Corp., Minneapolis, MN), 1974.
30. "Guartec CIP," Chemicals for the Paper Industry Bulletin, General Mills Chemicals, Inc. (now Henkel Corp., Minneapolis, MN), (A27089).
31. "Guartec in Corrugating Medium," Technical Information G-101a, General Mills Chemicals, Inc. (now Henkel Corp., Minneapolis, MN), Sept. 1965.
32. K. E. Lowe, "Secondary Fiber Requirements for Semichemical Corrugating Medium," in "Forming and Recycling of Paper Making Fibers," Western Michigan University, Kalamazoo, Jan. 1971, pp. 102-103.
33. Z. Chlubek, *Przegl. Papier.*, 22(12): 389-90 (1966).
34. "Cationic Guar Derivatives for Paper Mills," Gendriv—Applications Bulletin GE-4, General Mills Chemicals, Inc. (now Henkel Corp., Minneapolis, MN), Aug. 1974.
35. "Cationic Guar Derivatives," Gendriv Chemical Product Bulletin GE-3, General Mills Chemicals, Inc. (now Henkel Corp., Minneapolis, MN), Aug. 1974.

4

Cationic Starch as a Wet End Additive

Donald S. Greif and Lawrence A. Gaspar

INTRODUCTION

Overview of Applications

Cationic starch finds wide usage in the paper industry as a wet end additive, as a size press starch, and as a binder in pigmented coatings. Its utility in the industry is based upon its ionic ability to be attracted to and to react with the anionic sites of cellulose, pigments, and other anionic materials in the papermaking furnish. While its size press and coating binder uses are important, this chapter concentrates on the use of cationic starch as an internal additive to the papermaking process.

When produced as an internal strength additive, the starch backbone is usually kept in its highest molecular weight state and may be modified only to control dispersibility, ionic charge density, and balance of charges. As used in the wet end today, cationic starch has found usefulness in improving paper machine runnability and efficiency, lowering furnish costs, and improving paper quality and end use paper properties. These aspects are discussed later in the chapter. Cationic starches of today can vary widely in properties and may vary in design for specific papermaking conditions, as well as specific desired improvements. Special cationics are made, for example, for alkaline papermaking, while others are made for acid side papermaking for either moderate or high total acidity conditions. Some products may be designed to provide strength only, while others are also designed to maximize the retention of the fine components of a sheet and lock them in to prevent printing press pick outs. Initially, cationic starch found its largest acceptance in fine paper as a dual strength and retention additive. Later it found expanded uses in unbleached grades of kraft and groundwood furnishes. It is today utilized in combination with almost all papermaking fibers, including reclaimed fibers and post-consumer waste. Cationic starch has also gained acceptance in such special applications as emulsification and retention of highly reactive internal sizing agents, such as ketene dimers and alkenyl succinic anhydrides.

Historical Background

The use of starch as an additive to impart greater dry strength has been practiced for many years and may be nearly as old as papermaking itself.

Starches from many sources—including corn, tapioca, potato, and wheat—have been used, depending on availability and economics. These starches have generally been of the unmodified type. In most cases the starch has been added in cooked or gelatinized form. In some instances, especially on heavy board grades, it has been advantageous to add the starch in uncooked form, relying on gelatinization in situ during the drying operation. Resembling hydrated cellulose in chemical structure and behavior, starch has, over these many years, allowed the papermaker to increase the strength of his paper with less refining.

Since about 1940, attempts have been made to improve the effectiveness and ease of handling starch. Pregelatinized starches which could be added dry to the beater and would swell in contact with cold water were introduced. This was helpful to the papermaker in the days when pulp preparation was entirely a batch system and the need for rapid response to control quality changes was not as critical as it is today. In attempts to improve the retention of starch, which has little affinity for fiber in the wet state, approaches have been taken wherein the starch is gelatinized in the presence of caustic, borax, or soaps. The alkaline complex formed precipitates upon contact with alum in the stock system. Converted starches have also been used to a limited extent and have been beneficial in some instances. Unmodified starches blended or in conjunction with natural gums have been employed fairly extensively in unbleached kraft grades.

The introduction of chemically modified ionic starches in 1954 represented a major advancement. These were corn starch derivatives containing cationic groups as part of the parent molecule. Modification of the starch involved the introduction of a tertiary amine hydrochloride group through an ether linkage. In water dispersions, the cationic derivative ionizes to yield a chloride ion and a positively charged starch structure which was found to migrate readily to the cathode in an electrophoresis cell.

The invention of cationic starch was a result of a need to improve upon the traditional use of alum and high levels of unmodified starch to achieve strength. Early laboratory evaluations actually produced overflocculated handsheets that tested low for mullen. This powerful interaction of cationic starch with cellulose fibers eventually came to be controlled, and mullen improvements with cationic starch were achieved readily.

From 1955 to the present time there has been considerable developmental effort connected with applications of cationic starch. In addition to providing increased quality and production speeds, cationic starch has played an integral part in the development of new paper grades. The rapid growth rate of high-speed, quality web offset printing was assisted by the internal bond contribution of cationic starch, and this grade should continue to utilize large quantities of cationic starch. The estimated consumption of cationic starches by U.S. papermakers from 1966 to 1977 is shown in Table 4.1.

Table 4.1
Consumption of Cationic Starches by U.S. Papermakers—Estimated

	Estimated Cationic Starch Consumption (Millions-Pounds)		
Year	Total	Wet End	Size Press
1966	42	42	0
1967	36	36	0
1968	45	42	3
1969	53	48	5
1970	65	50	15
1971	81	57	24
1972	87	63	24
1973	101	77	24
1974	118	92	26
1975	101	77	24
1976	137	105	32
1977	132	105	27

Over the last 25 years, various types of cationic starch have been developed and include both tertiary and quaternary derivatives of corn; potato; and, to a minor extent, tapioca and wheat starch.

New cationic starches are under constant development to satisfy the changing needs of the paper industry. More efficient products which increase paper strength properties, maintain performance under a variety of paper mill conditions, and increase strength of retention under alkaline or high-acidity conditions are in use today.

MANUFACTURE

Chemical Nature of Starch

Starch belongs to the class of organic compounds called carbohydrates and is composed of carbon, hydrogen, and oxygen in the ratio $C_6H_{10}O_5$. These atoms are organized into a simple sugar molecule, D-glucose, or dextrose as it is known commercially.

The glucose units are connected through an oxygen atom joining carbon atom number one of one glucose unit to carbon atom number four of the next glucose unit. Thus, starch is considered a condensation polymer of glucose formed by elimination of water between glucose units.

It is evident from the structure of starch that modification can take place at the hydroxyl groups (–OH) through reactions such as oxidation, esterification, or ether formation. The linkages between the glucose units in the chain are recognized as acetal linkages by the organic chemist and thus subject to scission (depolymerization) by acid hydrolysis.

Commercial starches contain very small amounts of proteins, fats, and phosphates, depending on the variety.

Physical Nature of Starch

Starch is derived from two main plant origins, cereal and root, such as corn, potato, and wheat. The isolated corn starch, which is the principal source of raw starch, exists in granular form with a size of about 25 microns. Two fractions exist in the granule, amylose straight chain and amylopectin branched chain, with about 72% amylopectin. Molecular weight estimates of 40,000 to 340,000 for amylose and as high as 75 to 100 million for amylopectin are reported. Each molecular form has its own specific properties.

The main differences among the various starch species lie in the physical characteristics of the starch granules, as well as the relative lengths and arrangements of the molecular chains. A starch can be readily identified by microscopic examination of the granules. However, granule characteristics in themselves do not determine the differences in properties that are important to end uses in the papermaking process.

One concept of molecular arrangement with the granule is that of a parallel closely packed and intertwined formation of chain and branched molecules laid down in "onion skin" layers. This is an extremely stable arrangement resistant to physical disruption in the dry state. However, water penetrates the intermolecular areas on heating and causes the granule to swell. At near the boiling point, the granule disintegrates into a more or less completely dispersed, colloidal suspension.

Similarly, the starch granule, too, can be penetrated by chemicals in a water suspension. The multitude of reactions which starch undergoes can yield products with chemical or physical characteristics quite unrelated to the raw or pearl starch.

Granular Reactions

A large number of reactions for the preparation of cationic starches have been cited in the literature. Most of these are commonly carried out in aqueous slurries using alkali and a cationic nitrogen-containing reagent. These reagents include a tertiary amine with a reactive group, such as beta-haloalkyl, 2, 3-epoxypropyl, or 3-chloro-2-hydroxypropyl groups. The resulting starch product is finished at acidic pH in the form of a tertiary amine salt.

Additional reagents used in preparing amine derivatives of starch follow: 2, 3-epoxypropyltrimethylammonium chloride, (4-chloro-2-butenyl) trimethylammonium chloride, (chloropropyl) trimethylammonium chloride, diethyl (beta-chloroethyl) ammonium chloride, diethylaminoethyl methacrylate, and cyanamide.

The derivatized granular starch slurry can be concentrated in a centrifuge or similar device prior to drying and packaging for shipment.

These granular reactions are carefully controlled to prevent any crosslinking that would impair subsequent dispersion of the granules when cooked. The reacted granules are carefully washed to eliminate the by-products of the reaction, some of which could interfere with the papermaking process or purity of the final paper product.

Dispersed (or Cooked) Starch Reactions

Instead of maintaining an intact granule, the starch slurry can be heated with or without a catalyst to rupture the granules. The dispersed starch can be treated further with alkali reagents similar to those used in granular reactions. In practice, the cationic starch would have to be shipped in liquid form or be prepared at the point of use. To dry the dispersed starch would be prohibitive in terms of costs for most applications.

Included in this category is the in situ preparation of cationic starches. When these starches are prepared at the point of use, the process requires careful regulation to ensure quality and uniformity. In practice, wet end starches are prepared at low solids contents (i.e., 2-4% by weight) because of the resultant viscosity at higher concentrations. Additionally, one has to contend with any by-products of the reaction that remain because of the lack of washing.

Instead of a dispersed starch, it is also feasible to prepare a CWS (cold water swelling) cationic starch. These starches are prepared by granular reaction, washed, and cooked on heated drums or flash dryers to disrupt and burst the granules. The resultant product is water soluble.

PREPARATION FOR USE

General

In the preparation of the cationic starch for addition to the wet end, it is essential to achieve full dispersion of all the ionic molecules into a colloidal state. Temperature-time relationships, which depend on the method of cooking, must be followed.

Cationic starch for wet end use can be cooked at any concentration below 4%. However, addition to the stock must be made at 1% or less for complete dispersion in the stock stream. It must be remembered that the high substantivity to fiber will result in nonuniform absorption if either the starch or stock concentrations, or both, are too high. Cook out can be checked by microscopic examination of a sample of cooked paste stained with iodine. Any uncooked swelled granules are easily discernible. In such a state, this portion of the starch contributes little toward retention, improvement of formation, or strength.

Batch Cooking

Cationic starch may be prepared by slurrying in water and cooking at up to 4% solids at over 205°F for a minimum of 20 minutes. It should be diluted to 1% solids or less prior to addition to the machine.

Continuous Cooking

Continuous cooking makes it possible to deliver high output of low solids solutions without the necessity of large storage tanks for the addition of cationic starches to the wet end of a paper machine.

Starch slurry of known concentration (usually 25-35% solids) is metered to a cooking chamber by a positive displacement pump. The starch may be diluted with predetermined amounts of water (4-6% solids) before or after cooking, or both. As the starch enters the chamber, it is immediately contacted by live steam in such a manner that intimate mixing occurs. The steam supplies both heat and turbulence to the system. As a result, the starch is instantaneously gelatinized.

The actual amount of after cooking dilution water supplied is dependent upon both the equipment and processing technique employed and the desired end product. With the dilution feature, it is possible to slurry high solids batches and control the dilution to suit the desired end product. When going directly to the paper machine, the cationic starch is metered at 2-4% solids and further diluted after metering so that it enters the machine system at less than 1% solids.

GENERAL CHARACTERISTICS

Types of Cationic Starch

The types of cationic starch can be classified by their capabilities to interact with the anionic fibers and fillers. Broadly speaking, the amount and type of substitution determines the cationic charge on the dispersed starch units. Figure 4.1 illustrates the difference between native corn starch and an acidified tertiary amine derivative. The zeta potential is plotted as a function of pH. A primary amine imparts a lower cationic charge at any particular pH, while a quaternary amine would be slightly higher. The cationic starch has sufficient ionic charge to combine with the anionic components in the furnish, in the typical papermaking pH range of 4-9.

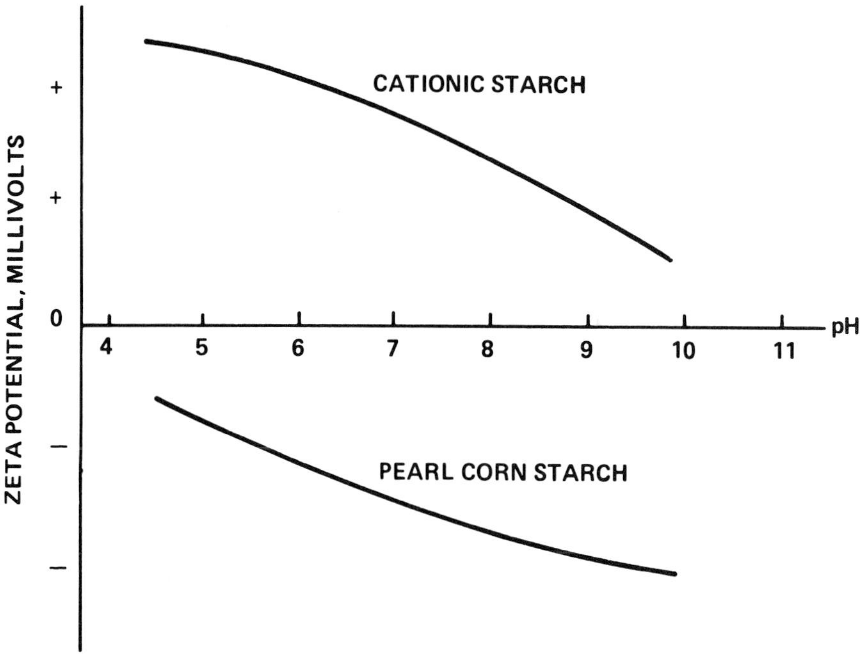

Fig. 4.1. Effect of pH on zeta potential of cationic starch and native corn starch.

In a paper mill system, the interaction of a cationic starch with the fiber, filler, and other components is influenced by the other ionic species, too. The paper mill wet end operating conditions can vary widely from grade to grade and from mill to mill.

Very early in the use of cationic starches, it was recognized that acidity from alum in moderate to high levels can change the requirements for a cationic starch to those of an amphoteric starch. The highly cationic alum at low pH competes for fiber and filler adsorption sites with the straight cationic starch. Proper balancing of the anionic and cationic substitution on the starch backbone and control of the overall charge density are vital to the effectiveness of the total system.

Adsorption Capacity of Papermaking System

The attachment of cationic starch to cellulosic fibers is quite tenacious. The maximum amount of cationic starch adsorbed is directly proportional to the active surface area of the fiber, as well as its relative degree of anionic charge.

As early as 1963, it was reported that the relationship between active surface area and maximum adsorption of cationic starch is very precise.

It has been shown that cationic starch essentially is adsorbed irreversibly. Furthermore, 90% of that added is instantaneously taken up by a variety of fibers—such as unbleached kraft, bleached kraft, alpha cellulose, bleached sulfite, and NSSC—with the remainder adsorbed within 15 minutes.

These fibers have a large capacity for cationic starch that ranges from a low of 5.3% to a high in excess of 7.0%. The fiber capacity is influenced by fiber refining, with greater adsorption at higher refining, due to the greater number of sites on the cellulose available for interaction.

The dramatic improvement in starch retention in the paper is reflected by a reduction in BOD and ppm starch concentration in the headbox and white water, as shown in Figure 4.2.

Retention

Introduction. The need for better retention has been stimulated in recent years by the trend toward higher ash contents; more efficient, cleaner operating wet end systems; and increased mill recycle of water (closure).

Retention may be reduced to two major effects, physical-mechanical and chemical. Physical-mechanical effects are influential with slow manufacturing speeds, highly refined fibers, and heavier basis weight grades. Conversely, high speed machines, lightweight paper, and moderately refined fibers rely more on flocculation effects to optimize retention.

Paper furnishes contain fine solids from a wide variety of sources. Examples include fillers for optical effects and fragments of fibers, termed "fiber fines." Organic by-products from pulping and bleaching are introduced into the papermaking furnish through system upsets or as a result of the economics of pulp washing or demand.

Filler Retention. The addition level that is more beneficial *for retention only* ranges from 0.25 to 0.50% in practice. Figure 4.3 illustrates the effect of the addition level for a cationic starch on titanium dioxide retention. Retention peaks at 0.50% addition. It is important to recall that more filler in the paper can have a distinct effect on strength; filler typically has a de-bonding effect. Additional quantities of a retention aid can be desirable when the retention aid also increases or maintains strength.

Cationic Starch. Four important points should be emphasized:

1. Retention of the starch itself;
2. Retention of other additives, fillers, and fiber by the starch;
3. Permanence of the bond between starch and the filler or fiber, particularly if repulped; and
4. Complete cook out of starch for maximum benefit.

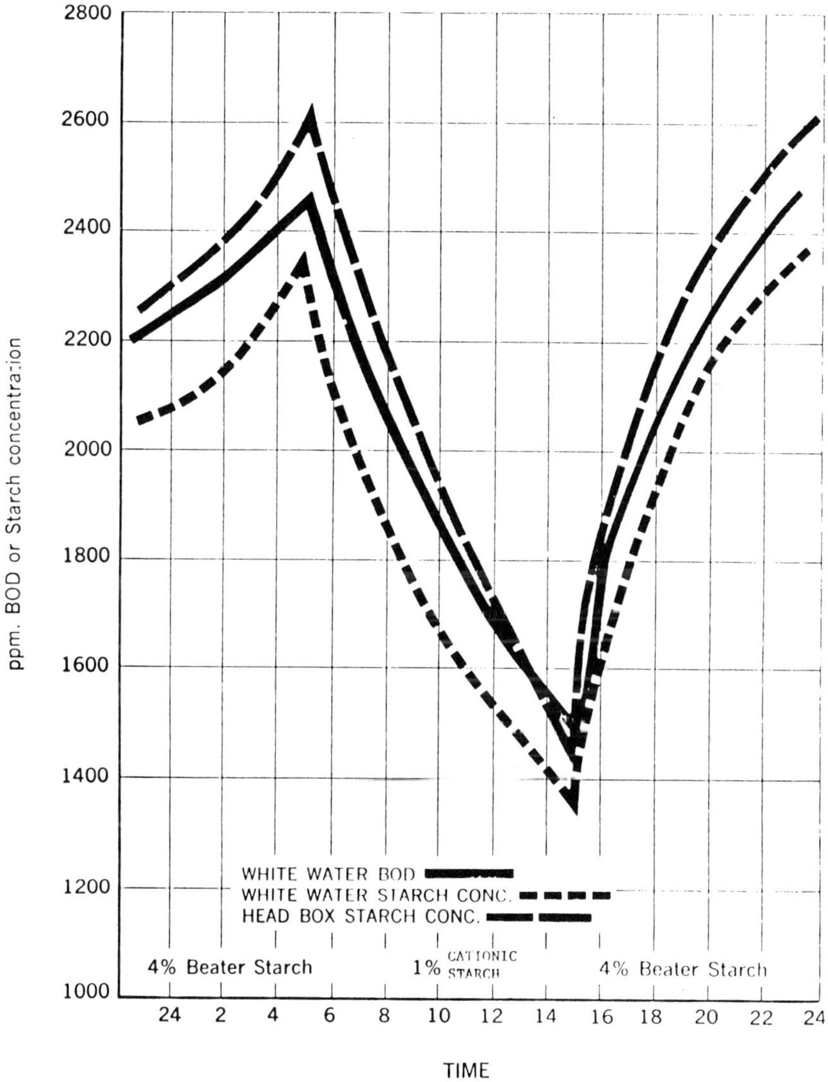

Fig. 4.2. White water starch concentration and BOD with beater starch versus cationic starch.

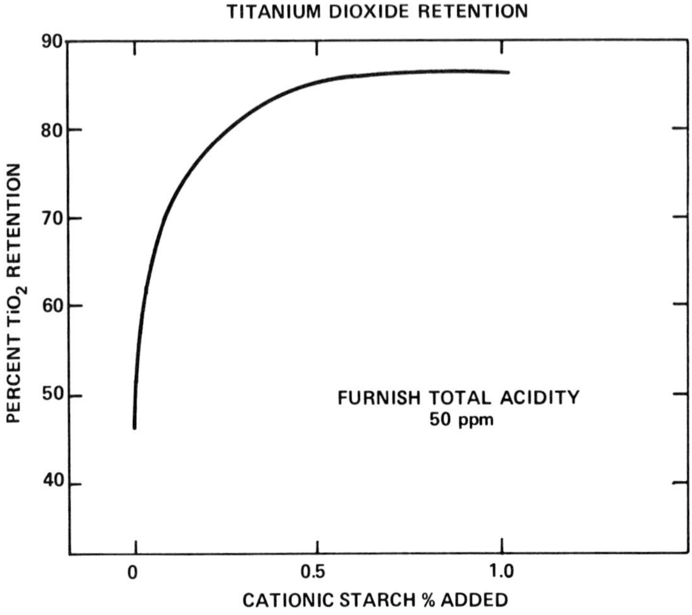

Fig. 4.3. One pass titanium retention versus rate of addition.

Pearl starch, non-ionic derivatives, and various anionic starches have relatively little inherent affinity for fibers or fillers. Retention of the starch depends more on mechanical-physical forces, and this is typically reflected in 40-50% of the added amount. Combinations with alum or even synthetic products have reportedly shown slight increases.

Oxidized starch containing anionic carboxyl groups adversely affects the retention of titanium dioxide. Conversely, cationic starch is effective in retaining titanium dioxide.

The adsorption and retention of cationic starch is essentially irreversible and complete in typical levels employed in papermaking, up to 4%, and these bond forces are strong enough to withstand processing and reuse of paper mill trim or broke. The process of adsorption is quite rapid and effective with pulps in paper. Alum competes with cationic starch for adsorption sites.

Strength

The development of strength is the major reason for the use of cationic starch today. It develops its strength characteristics in several ways. First, the cationic starch electrostatically bonds with the anionic sites on long fibers, fiber fines, and typical mineral fillers. As a result of the ionic bonding, the cationic starch is brought to intimate contact at the fiber surface. There the additional hydrogen bonding forces reinforce the strength of the cellulosic bonds.

The strength contribution of cationic starch is the joining together of fiber fines and long fibers into a cohesive network. Direct benefits on the paper machine result from better runnability and fewer wet end breaks. The stronger bonds in the wet state also mean a cleaner wet end operation. The more firmly held filler and fines in the wet web remain there, rather than deposit on press rolls, fill felts, or close up the holes in vacuum boxes and couch rolls.

This bonding of fiber fines contributes to improvements in mullen, internal bond (z-directional tensile), and fold endurance. Improvements in tear can be achieved indirectly through less refining and through using the cationic starch in place of the refining for maintaining other physicals.

Surface strength properties of paper are improved with cationic starch for many end uses. Fiber bonding increases are reflected in better IGT pick, wax pick, and scuff resistance. Additionally, there is less linting and dusting, and this is particularly beneficial during printing operations.

Mineral filler contact and distribution in paper is of considerable interest from both overall strength and uniformity of strength bases. It has been shown on many fourdriniers making fine paper that there is a higher filler content on the top or "felt" side of paper. The filler acts to reduce bonding and thus creates a difference in strength throughout the thickness of the paper. Cationic starch provides a double benefit. First, it retains more filler, and, additionally, there is a strong tendency for the filler to be distributed uniformly. Cationic starch not only acts to retain the filler in the web, but it also bonds the filler more securely.

Value of Cationic Starch in Reclaimed Paper or Mill Broke

It is claimed that the cationic starch in reclaimed paper or broke is retained in the reconstituted slush stock. This not only contributes to dry strength of the paper in which the broke is used, but it also prevents the release of starch in the white water.

When cationic starch is used at the size press, the effect is more pronounced because the amount involved is larger. It is not uncommon to find as much as 300 pounds of soluble starch/ton in the headbox of a fine paper machine. This concentration of starch in the white water contributes to a high BOD in the effluent. A high percentage of cationic size press and coating starches is retained with the repulped fiber and pigment and is thereby retained on the wire, thus reducing the soluble starch in the white water.

Effect on Other Additives

The better retention and bonding efficiency of cationic starch for fiber fines and other additives contributes to their more efficient use. Additives such as size adsorb readily on the large surface area of the fiber fines which, when retained by cationic starch, produce increased effectiveness.

More efficient sizing (i.e., rosin), dye utilization, and wet strength have been achieved. As an example, rosin size has been reduced in uncoated printing papers. An average savings of 3.5 lb/ton (30% reduction) has been reported in a mill while utilizing 4-6 lb/ton of cationic starch.

General Effects on Printability

Improved printability may be one of the greatest benefits resulting from the use of cationic starch. Better print fidelity and sharper images have been attained on offset (both sheet and web fed), letterpress, and rotogravure printing presses.

Equally important have been the cleaner operations on the presses, particularly with offset printing. Cationic starch in both uncoated and coated grades of paper has contributed to achieving high quality reproductions. Additionally, cationic starch eliminates pick outs due to weakly bonded filler, fiber fines, and hardwood vessel segments.

The advent of web offset printing presses has created a challenge to the paper industry to control blister, primarily in coated papers. Cationic starch has assisted with internal bond (Scott Bond) increases.

PAPER MACHINE APPLICATIONS

Selection of Addition Point

The interaction of cationic starch with fiber is rapid. Because of the speed and nature of the bonding action between cationic starch and the furnish, it is important to exercise care in obtaining uniform distribution. Investigation has shown that when ideally uniform distribution of adsorbed cationic starch is obtained, as opposed to nonuniform distribution, the effectiveness may be improved by as much as 50%. These observations form the basis for the following recommendations:

1. Cationic starch should generally be added at a maximum of 1% solids.
2. Normally the cationic starch is added as close to the headbox as possible. This minimizes excess shear that usually impairs retention.

3. There are several choices for achieving the optimum dry strength response. It has been added effectively near the headbox, at the stuffbox, or at the machine chest. The determination of the best point is based on analysis of each system. Generally, the longer contact time with the furnish maximizes adsorption and hence the strength improvements.

4. Where several types of fibers are used, i.e., hardwood and softwood, it may be important to add the cationic starch to only one. This requires analysis for determining the best choice.

5. Part of the cationic starch for retention and the remainder for strength may be added at two separate points. As noted in the retention section, from 0.25 to 0.50% cationic starch should be added close to the headbox.

6. Proper neutralization of highly anionic furnish components is important. Highly anionic furnish components can adsorb the cationic starch and thus make it unavailable for strength or retention.

7. Cationic materials like alum can compete with some cationic starches for adsorption sites. Thus, it is important to add cationic starch at a different point. Adding cationic starch after alum, and in some cases before alum, has been successful.

Effect of Acidity

Alum containing systems have been utilized in papermaking for many years, and alum is the primary source of acidity in most instances. Thus, the chemistry of alum becomes an integral part of the use of any product as a retention and strength additive for fiber, fines, and fillers and is recommended for study.

Cationic starch generally is effective in low to moderate acidity systems. Since other cationic species such as alum compete with straight cationic starch, the total acidity should be kept to a moderate level. This aids maximum adsorption and leads to the best results. The influence of total acidity on cationic starch is shown in Figure 4.4. Cationic starch can be modified for maximizing performance under varying alum conditions.

APPLICATION TO SPECIFIC GRADES

Fine Paper

Uncoated. Cationic starch effectively minimizes hardwood vessel segment picking on uncoated printing paper. Physical tests are listed in Table 4.2.

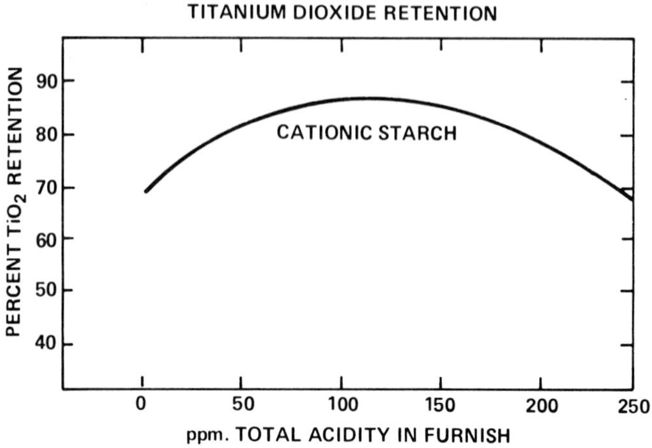

Fig. 4.4. Influence of total acidity on one pass titanium retention with 0.25% cationic starch added.

Table 4.2
Vessel Segment Picking Reduction with Cationic Starch

	Cationic Starch Addn.		
Test Average P	0	4#/T	8#/T
Opacity	80.7	81.5	82.0
Mullen	22	24	23
Felt Side Wax	13	13	14
Print Test-W/F[a]	7/22+	5/14	2/6
IGT W/F	230/106	265/185	302/200

[a]Print Test: Simulated offset printing tests expressed in number of vessel segment picks per unit surface area.

An example of cost reduction is shown for uncoated printing papers:

Rosin size added, 1 lb/T: range 0 to 7 lb reduction, average reduction 3.5 lb/T
Cationic starch added: 3.9 to 6.8 lb/T
Ash increase (clay): range 0.5 to 4% – average – 3%
Change in mullen/pick: none, maintained at higher ash content
Dollars/T saving: range 0 to \$12.30, average saving– \$3.50/T
% production increase: range 0 to 26%, average– 8% attributed to better drying

Cationic starch at 7 lb/ton was evaluated against 0.8 lb/ton synthetic retention aid in an attempt to upgrade surface strength of a 46 pound offset paper as measured by wax pick. During the run, improvements were noted in wax pick and ink penetration (internal sizing) at higher ash content.

The performance capabilities of cationic starch for allowing greater utilization of lower strength fiber in a fine paper grade is illustrated in Table 4.3.

Table 4.3
Effect of Cationic Starch on Strength With Increased Bleached Hardwood

	Scott Bond	IGT Pick (5-50-B)	Wax Pick	% Ash
		W/F	W/F	
32% SWD[a]-No Cationic Starch	128	284/177	14.7/10	10.2
20#/Ton Cationic Starch	153	310/238	16 /13	10.6
Reduced Refining	127	218/167	15 /11.5	12.0
23% SWD-20#/T Cationic Starch	148	320/245	16 /14/5	12.2

[a]SWD – Bleached Softwood

Coated. The primary function of cationic starch in coated printing grades is to improve properties of the base sheet so that not only is base stock pick strength of the base sheet improved, but final surface pick after coating is increased. This is due to the stronger surface fiber bonding in the base sheet, which increases the interface adhesion of the coating to the base paper.

Data from a mill run with cationic starch on a letterpress coated grade is shown in Table 4.4. The run was made on a 50 pound machine coated letterpress sheet containing 13-15 pounds of coating per ream.

Table 4.4
Coated Paper Strength Improvement with Cationic Starch

	CONTROL	CATIONIC STARCH	
		10#/T	14#/T[a]
Basis Weight	50#	52#	50.5#
Opacity	83	84.2	84.8
Mullen	20	23	19
Wax Pick	W7/7F	W9/8F	W10/8F
Total Ash	30%	34.6%	35.4%
Base Stock Ash	9%	10.2%	14.9%

[a]Clay increased

Figure 4.5 shows a photostat of offset printing proof tests of on-machine coated papers. This test procedure duplicates commercial printing results to a very high degree. It reflects all of the properties mentioned above, particularly the picking out of hardwood vessel segments. These photostats very vividly show the differences in offset ink pick resistance of paper.

Neutral and Alkaline Papermaking with Calcium Carbonate. The change from acidic to alkaline conditions influences the chemistry of almost all additives, including dyes, sizing, and fillers. As a filler, calcium carbonate can be utilized for brightness at lower cost. Thus, retention and strength for calcium carbonate becomes essential and satisfies any change calcium carbonate contributes to the colloidal change in the furnish as well.

In a system with a synthetic sizing agent, cationic starch has boosted filler retention and additionally retained the synthetic sizing agent. Results obtained in a lightweight bond grade illustrate the retention improvement of a clay and calcium carbonate filler. At pH 7.2, 15 pounds of cationic starch increased the first pass ash retention from 22 to 51%.

Depending on the system, small quantities of alum can be used for a synergistic effect in combination with cationic starch. For an uncoated offset grade the following, in Table 4.5, is typical:

Table 4.5
Effect of Cationic Starch on Calcium Carbonate and Clay Fillers Retention

Fiber	—	Bleached softwood/hardwood (30/70) kraft
Filler	—	Clay/calcium carbonate 13%/7%
pH	—	7.4
Ash Content at Reel	—	14-15%

	Additive	First Pass Ash Retention
a)	0.25% Alum	26%
b)	0.6% Cationic Starch	54%
c)	Both a) and b)	68%

In some systems the use of a synthetic retention aid can be beneficial. However, the use of this retention aid to the exclusion of the cationic starch must be examined from a strength viewpoint, as well as other benefits noted during operation of a paper machine. Though the synthetic retention aid retains the filler and fines, the bonding of the filler and fines within the sheet typically requires a strength additive or increased refining.

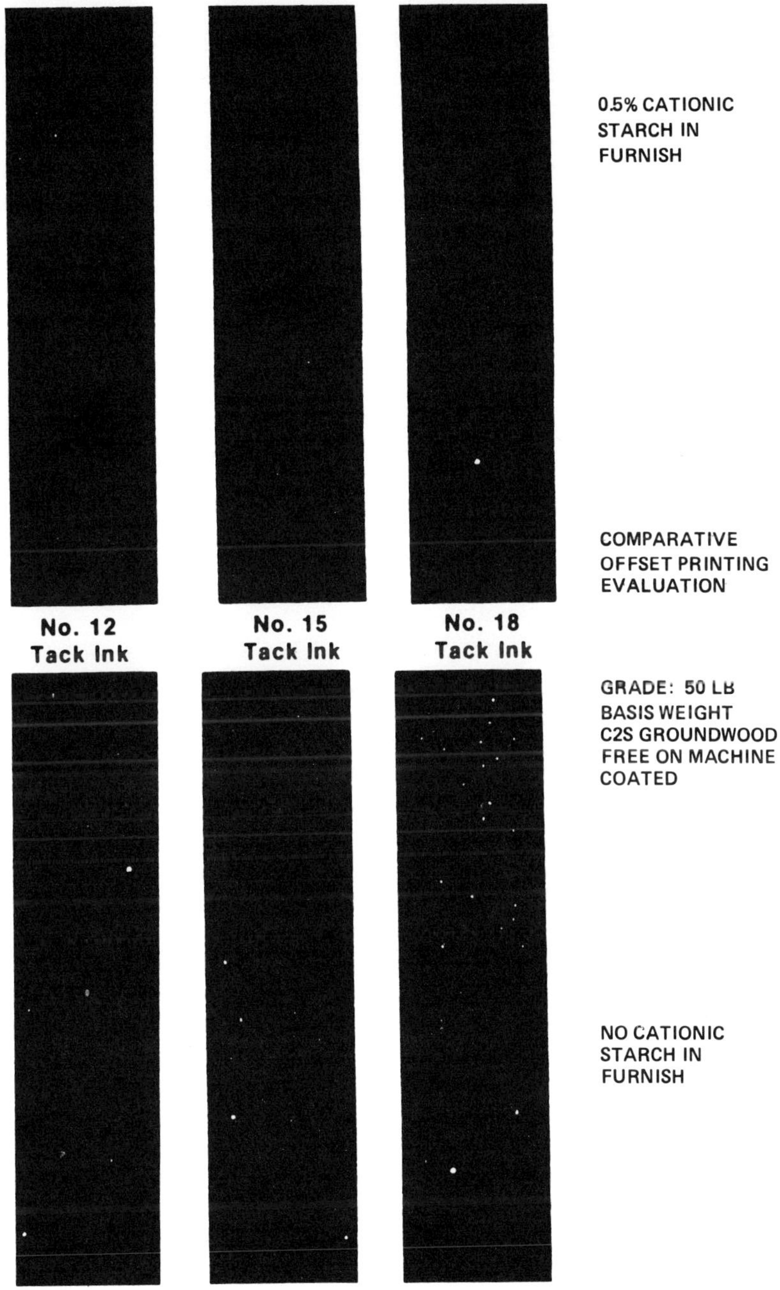

Fig. 4.5. Comparative offset printing evaluation.

Unbleached Kraft

Bag Stock. Cationic starch has proven effective for increasing strength factors, such as mullen and tensile, at average concentrations ranging from 0.2%-2.0%, based on dry fiber weight of bleached or unbleached kraft grades. On bag paper, tear increases have been realized when refining is altered and cationic starch is used to maintain tensile and mullen with less refining. A more "squared" sheet is obtained with greater equalization of machine and cross-machine tensile. Of particular interest where unbleached kraft grades are concerned is the fact that stock with cationic starch drains freely on the wire, in many cases enabling faster machine speeds and greater production while maintaining strength.

Linerboard. Cationic starches have been employed in linerboard furnishes as dry strength aids. The use of cationic starch adds considerable operating flexibility in linerboard manufacture. Improvements in mullen, particularly, and other important dry strength tests create several options to consider for optimizing profitability.

Figures 4.6 and 4.7 further depict the mullen increase with a cationic starch, which can enhance the full range of test linerboard basis weights (i.e., 42-90 $lb/1000\ ft^2$).

Cationic starch increases mullen and other strength properties of recycled board furnish. The exact response is influenced by the proportion of kraft and corrugating medium fibers. Generally, the strength improvement is close to that of a 100% virgin fiber system.

Fines retention in unbleached kraft can be a valuable operation improvement and is needed to assist in strength development, to optimize yield, to increase saveall operation, and to reduce sewer losses.

The use of cationic starch has demonstrated improved fines retention, as shown in Table 4.6.

Table 4.6
Effect of Cationic Starch on First Pass Retention in Linerboard

	First Pass Retention	Tray Water Consistency
Before	73	.16
Cationic Starch, 0.75% Addition	80	.12

Groundwood-Containing Grades

General. Newsprint, uncoated printing papers (or specialties), and coated publication grades contain groundwood contents ranging from 85-40%. Operating demands and printing press performance (freedom from picking and web

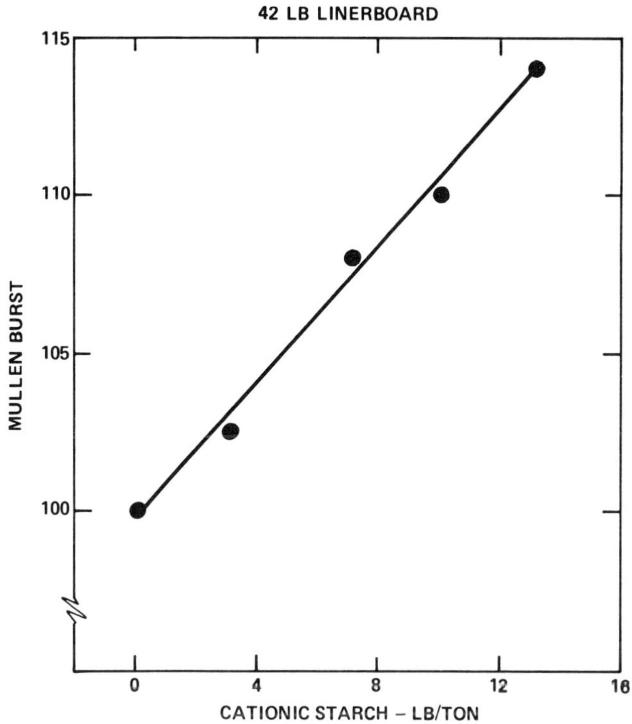

Fig. 4.6. Effect of cationic starch on mullen increase in 42 lb linerboard.

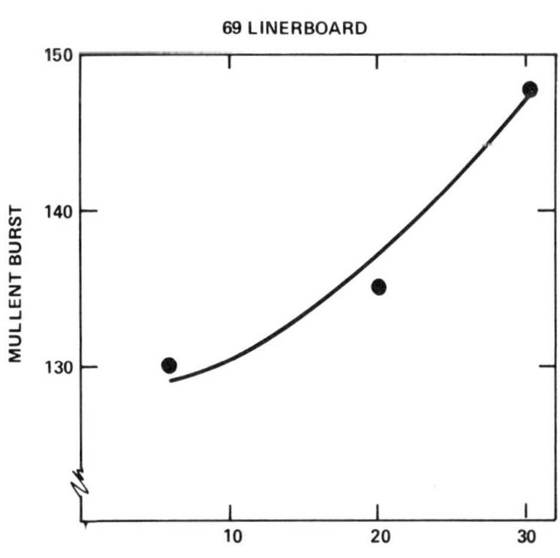

Fig. 4.7. Mullen improvement with cationic starch in 69 lb linerboard.

breaks) require that the fines generated from the groundwood portion be retained and securely bonded.

The process of gaining fines retention along with the needed strength can be met with a combination retention and strength additive, i.e., cationic starch. The choice of cationic starch is influenced by stock conditions, including pH, total acidity, and final end use requirements.

Cationic starch in high groundwood content papers has been studied for optimization of strength, including retention of the cationic starch. The addition of alum resulted in a drop of the total amount of cationic starch that could be adsorbed onto the fiber from 7% to 4%. It has been suggested that the alum added prior to starch occupies fiber sites that otherwise would be available for starch attachment.

More importantly, when the cationic starch was added to the groundwood first, maximum starch retention was achieved. Furthermore, the preferential treatment of the groundwood resulted in the best strength response. Thus, a good correlation between retention of the cationic starch on the groundwood and Scott Bond was noted.

Publication Coated. Typical examples of experiences with cationic starch in this grade are as follows:

1. A mill making supercalendered coated web offset from a furnish of 50% groundwood and 50% kraft added 15 lb/ton of cationic starch to the machine chest. A blister test showed a significant reduction in the tendency to blister, compared with a control made with retention aid plus pearl starch. The reduced blistering allowed the mill to precoat the sheet for better overall coating properties. Even with this precoat, blistering was controlled. Blistering on coated sheets is caused by entrapped moisture in the sheet where heat set inks are used in the printing process. The water turns to vapor which expands and causes blisters. By making a stronger base sheet, blistering can be reduced, and it is thus directly related to internal sheet strength.
2. Typical improvements obtained with cationic starch on supercalendered 40 pound coated web offset over a 30 day period are shown in Table 4.7.

Table 4.7
Web Offset Paper Strength Improvement with Cationic Starch

		IGT Pick		Wax Pick	
Cationic Starch Level	Scott Bond	Wire	Felt	Felt	Wire
0	113	155	185	5	7
5 lb/ton	117	165	200	6	7
10 lb/ton	160	220	240	7	7
15 lb/ton	195	225	240	7	8

To achieve these desirable strength and printing properties, however, careful attention must be given to the point of addition of the cationic starch. Early mill trials on groundwood specialties showed that addition of cationic starch at the more conventional points, such as the headbox, could cause excessive reactivity between the starch and the furnish. This may result in reduced brightness because of excessive retention of low-brightness groundwood fines or picking at the wet presses caused by pitch retention. Poor formation could also be caused by overflocculation.

By selecting points of addition further back in the system, the papermaker can avoid the problems of excessive flocculation and retention and achieve a balance of sheet physical characteristics and operating conditions to fit the requirements for many groundwood papers. Generally, it has been found that the starch should be added to the suction side of the fan pump or earlier. Addition to the mixing box prior to a primary fan pump or the stuff box is also suitable. Under some circumstances, it is desirable to add the starch in the machine chest. Some papermakers add the starch to the groundwood before mixing with chemical pulp. One of these has used the cationic starch for some time to reduce dusting and linting. This mill uses between 1.0 and 1.5 percent of starch and adds it to the groundwood furnish only, prior to mixing with chemical fiber.

In another instance a "split" addition of the starch proved to be the answer to a printing problem. This mill was making a 40 pound web offset grade that was filled and supercalendered but not coated. Both the wire and the felt side were giving problems on five-color offset presses. The felt side problems were especially severe.

To correct this, 15 lb/ton of cationic starch was added to the suction side of the fan pump. In comparison with the previously used retention aid (a polyamide/polyamine polymer), wax pick was improved on both the calendered and the uncalendered stock. However, neither formulation gave completely acceptable press performance. The control made with a retention aid could not be run because of pick on both sides. The paper made with the cationic starch was acceptable on the wire side, but the felt side picking required frequent shutdowns. The point of addition for the cationic starch was changed to 5 lb/ton at the suction side of the fan pump and 10 lb/ton at the blending chest. As a result, the wax pick was increased from 4 to 10 on the felt side, and the wire side increased from 5 to 14. The grade could then be satisfactorily printed.

Newsprint. In offset newsprint, the use of cationic starch has resulted in improved performance on offset presses. Reduced linting and dusting on the press is of primary importance in offset news. This property is difficult to measure, and mills are often totally dependent on the printer's comments in evaluating their sheets.

To measure the effect of a cationic starch on lint levels, an offset newsprint mill varied the amount of starch used and then compared lint type, lint levels, and surface strength of the sheet. As the starch level increased from 7 lb/ton to 20 lb/ton, the amount of lint sharply decreased from 98 to 53 mg. In addition, the type of lint changed in character. At 7 lb/ton, the lint consisted of groundwood fiber bundles, long kraft fibers, and groundwood debris. At the 10 and 15 lb/ton level, the lint type had changed somewhat. At 20 lb/ton the only lint present was groundwood debris. The IGT test went from 57 to 72.

Using a visual rating, a mill making a 25 pound web offset grade found that cationic starch substantially decreased linting. This mill evaluated linting via a dry blanket multilith. After running 100 sheets, the blanket was visually inspected and rated from 0 (highest rating) to 7 (lowest rating). A rating of 4 or over was considered unacceptable.

Sheets made with 7 lb/ton of cationic starch were rated 1, whereas their normal sheet without additive ranked about 3. The starch was added at the stuff box at 2.5 percent consistency immediately after pulp blending.

Other mill results confirm that strength of straight newsprint can be substantially improved with cationic starch. With 7.5 lb/ton of cationic starch, the Scott Bond was improved from 107 to 127, and the IGT on the felt side went from 60 to 118. The wire side IGT remained at 210. The felt side is the side that causes most of the linting and dusting problems in printing, and the starch appears to upgrade preferentially this critical side.

These improvements in paper quality increase production efficiency on the paper machine by minimizing reject paper. Printing operations are also more efficient, because fewer web breaks and reduced linting permit longer runs between press shutdowns.

Board Grades

In grades where printability, interply bond, and finish are of importance, cationic starch affords increased pick strength and improved formation for better ink receptivity. Ply bonding is stronger through increased fiber adhesion. Additions of 0.25 to 1.0 percent to the top liner furnish are effective. Additions may be made to all plies to yield high strength as needed on bleached kraft container and sanitary board stock.

In a 120 lb-10 mil bleached raw stock board, the replacement of gum with 10 pounds of cationic starch permitted a hardwood increase from 55 to 60%, a Dennison wax pick from 6 to 7, and a speed increase from 740 to 760 ft/min.

OTHER FUNCTIONAL USES

An important, growing application of cationic starch in the paper industry is its use as a retention aid for synthetic sizes such as Fibran® and Aquapel®[1]. These materials are highly effective sizing agents and are commonly used to size paper on the alkaline side. This enables production of a stronger sheet with superior aging properties. Since alum is not required, as in rosin sizing, sizing agents of this type permit the reuse of coated broke containing calcium carbonate or the direct addition of alkaline fillers or other agents which require alkaline stock media. The size is a hydrophobic substance which is added to papermaking stock in the term of an aqueous emulsion. Cationic starch is employed in the preparation of this size emulsion. Frequently, additional cationic starch is added separately after addition of the size-cationic starch emulsion to the stock to yield further retention of the sizing agent and fillers and to promote further strength increases. It is important to note that cationic starch is effective in providing substantial improvements in retention and strength in alkaline stocks.

Surface sizing applications of 1-2% solids of a wet end cationic starch can be employed to reduce fuzz and improve finish on grades such as kraft liner and white envelope.

Machine applications indicate that the surface obtained with these small percentages is smoother and stronger than that which is secured with other commonly used starch surface sizes. Additionally, less dusting has been noted at the slitters.

Cationic starch has been used to increase substantially the retention of butadiene-styrene latices and various anionic resins added to the beater. Alum can generally be reduced or eliminated as a precipitant, and a stronger sheet can be obtained. Concentrations of cationic starch employed have ranged from 1-5%, based on the weight of dry fiber. Rosin sizing is frequently improved with cationic starch.

[1]FIBRAN® Trademark of National Starch & Chemical Corporation
AQUAPEL® Trademark of Hercules, Inc.

5

Dry Strength from Wet Strength Agents

C. S. Maxwell

INTRODUCTION

It was early recognized that the surface (size press, etc.) application of glue-formaldehyde and, later, urea formaldehyde gave dry strength as well as wet strength to papers, among which blueprint papers are a good example.

With the introduction of wet end added wet strength agents (melamine resin acid colloid, anionic and cationic urea formaldehyde resins, and polyamide polyamine-epichlorohydrin resins), the same general pattern was noted. At least under laboratory and carefully controlled mill tests, dry strength, as well as wet strength, was improved by the resins.

Since the preparation and use of these wet strength resins have been well covered in the literature (*1,2*), these topics are not described here.

Wet strength agents should not be considered as direct competitors to dry strength additives. If dry strength improvement alone is desired, wet strength resins complicate broke recovery unnecessarily and should not be used.

DRY TENSILE AND BURST FROM WET STRENGTH RESINS

The usual commercial wet strength resins—urea, melamine, and polyamide-polyamine-epichlorohydrin—improve the *dry* tensile and *dry* burst of paper as well as the *wet* tensile and burst. The effect of melamine formaldehyde resin on dry strength has probably been studied and documented in more detail than has the effect of the other resins; thus, most of the information is based on this resin.

From an early study of strength data from a large number of laboratory-prepared handsheets—some treated with melamine resin, some with the same furnish but without resin—emerged a pattern: the wet strength improvement in pounds per linear inch tended to be the same as the dry strength improvement (*3*).

If this dry strength-wet strength relationship is true with laboratory-prepared handsheets, there is a question as to why it was not as noticeable with machine-made papers. In some mill-made papers, a similar wet strength improvement-dry strength improvement had been noted; but, in many cases, little or no increase in dry strength occurred, although wet strength improvement was excellent.

One of the most striking examples of high dry strength and high wet strength improvement occurred in a carefully controlled run of currency-type paper using an all-rag furnish. Half of the highly beaten rag stock was run untreated; the other half, after treatment with 3% melamine resin. The resin-treated paper had over twice the dry strength of the untreated and a wet strength greater than the *dry* strength of the untreated, again about the same increase (lb/in.) wet or dry.

With increased production of commercially produced wet strength paper, the answer began to evolve. Melamine resin (as the acid colloid) tends to lower the freeness of paper stock, making it slower to drain on the forming wire. Paper machine crews tend to correct this slowness by backing off on refiners, which action in turn tends to lower tensile and burst while improving tear. Thus, the expected increase in tensile strength failed to show up, because it was unknowingly "tuned out" by the machine crew.

An interesting corollary was observed during the production and printing of high wet strength army map paper during World War II. One of the greatly desired attributes of such paper was dimensional stability during changes of humidity. Laboratory measurements of resin-treated and untreated handsheets showed no measurable effects attributable to the wet strength treatment. Yet the printers reported that the wet strength paper had much better dimensional stability than the non-wet strength. The answer proved to be the same as that for the missing dry strength improvement. When the resin-treated stock hit the forming wire, it was obviously slower than the untreated stock. Thus, the refiners were backed off to regain the desired drainage. Paper made from the less refined stock naturally had better dimensional stability. Only indirectly had the wet strength resin affected the dimensional stability.

The other two major classes of wet strength agents (urea formaldehyde resins, both anionic and cationic, and the polyamide-polyamine-epichlorohydrin resins) do not exhibit the same slowing effect that melamine resins do. The urea resins seem to have little effect, while the polyamide-polyamine-epichlorohydrin resins actually have a freeing effect on stocks.

DRY STRENGTH PROPERTIES—OTHER THAN TENSILE AND BURST

A negative aspect of the size press application of thermosetting resin for wet strength was the adverse effect on folding endurance. Under a reasonable cure with an acidic accelerator, the treated paper often had only 10% of the folding endurance of the untreated paper. Attempts to minimize this drop in folding endurance with various cures, various accelerators, and even plasticizers were complete failures. The wet strength and folding endurance always went in opposite directions.

With the discovery that a melamine resin colloid imparts excellent wet strength by wet end addition came the rather astonishing discovery that the folding endurance of the resin-treated papers was higher than that of comparable untreated papers. The difference in folding endurance was due to the method of application. When either melamine or urea resin is added to formed paper at the size press, the resulting paper has lower fold than the untreated; but, if either resin (suitably modified to make it substantive to the cellulose fibers) is applied to slush stock, the resulting paper has much improved fold. This decided increase in folding endurance is one of the reasons that melamine resin is used in U.S. and Canadian currencies. It makes money last longer before wearing out.

The internal bond or "Z" direction strength of paper pertains to how easily it splits or resists splitting. Wet strength resins improve the "Z" direction strength of fourdrinier sheets and improve the bonding between plies of multi-cylinder sheets. This improvement has had and still has commercial importance in a number of grades of paper. It has been of particular importance in currency paper. A stunt used on older currency was to "dog ear" a one-dollar bill and a twenty-dollar bill until a split started at the corner. This split could be extended the full area of the bills by simply pulling the two "ears" away from each other. (Of course, the two half-bills were recemented 1 to 20 and passed as 20's.) When the switch was made to melamine resin-treated currency, pulling on the two "ears" caused a run out to one side rather than splitting, because the "Z" direction strength was higher.

Another application was in coating raw stock. The resin treatment increased the surface and internal strength so that wax pick and other surface strength tests were materially improved. The use of wet strength resins for this use has dropped off, as other treatments are available to improve pick resistance without the unwanted wet strength.

Tear Resistance

The "wet tear" resistance of paper is appreciably enhanced by treatment with wet strength resins, but the normal "dry tear" is always reduced. This dry tear loss is attributed to the slight stiffening effect of the resins, which permits consecutive breaking of bonds rather than allowing a slight rearrangement of fibers that requires breaking of a number of bonds concurrently.

As mentioned earlier, this loss of tear resistance is often compensated for by backing off on refiners, which tends to bring the tear resistance up at the expense of tensile and/or burst.

MECHANISM OF DRY STRENGTH IMPROVEMENT

The basic mechanism for improvement of dry strength from wet strength resins or from dry strength additives is considered to be improved bonding between fibers.

In the case of such dry strength additives as starch, gums, and acrylamide polymers, it has been proposed that they provide supplementary bonds between fiber surfaces too far apart for normal hydrogen bonding forces to be effective. This hypothesis is supported by the observation that, as more dry strength is developed by beating, less strength improvement can be imparted by a dry strength additive.

The dry strength imparted by wet strength resins does not show this response. For example, a melamine resin improves the tensile strength of highly beaten stock to the same extent that it does lightly beaten stock. Obviously, the additional bonds furnished by the melamine resins enhance the dry strength in a manner quite different from the dry strength additive. The melamine resin bonds are developed by polymerization (condensation reaction) of the resin in the dryer section to form water insensitive bonds. These provide more bonding sites regardless of how far the dry strength has been developed by beating.

MISCELLANEOUS OPERATIONAL EFFECTS

While probably none of the commercially successful wet strength resins was designed for any particular effect on machine operation, it became apparent that some resins had a decided effect on the operating freeness of the stock. As described earlier in this chapter, melamine resin (acid colloid) has a decided slowing effect; urea resins have no significant effect; and the polyamide-polyamine-epichlorohydrin resin has a definite freeing effect. Thus, the slowing effect of melamine resin has been used to permit underrefining of stock to obtain a paper with less expansion-contraction due to humidity changes. The polyamide-polyamine-epichlorohydrin resin has been used in quantities below the level at which significant wet strength is obtained as a drainage aid on paper machines. This resin presumably causes at least partial coagulation of fine material into larger particles, which gives room for the water to drain more easily. Other derivatives of this resin have been formulated to optimize the drainage improvement properties and to minimize the wet strength effects.

In conclusion, the wet strength resins as a class are also excellent dry strength additives. Grades requiring wet strength get the added dry strength gratis; grades not requiring wet strength tend to be produced with dry strength additives, probably at lower cost and certainly with less broke recovery problems.

Literature Cited

1. K.W. Britt, ed., "Wet Strength in Paper and Paperboard," TAPPI Monograph Series No. 13, 1954.
2. J.P. Weidner, ed., "Wet Strength in Paper and Paperboard," TAPPI Monograph Series No. 29,1965.
3. D.J. Salley and A.F. Blockman, *Pap. Trade J.*, 125(1): 35 (1947).

6

Acrylamide-Based Polymers for Dry Strength Improvement of Paper

Walter F. Reynolds

INTRODUCTION

The commercial development of acrylamide monomer from acrylonitrile in the early 1950's made it possible to produce a new water soluble synthetic polymer with outstanding hydrogen bonding capabilities.

Polyacrylamide polymer is unique in the respect that a water soluble polymer results directly from the polymerization of the monomer. Like most synthetic polymers, it is readily manufactured in a wide range of molecular weights and in combination with other monomers.

Cationic or anionic polymers can be made readily, depending on the substituent incorporated in the polymer. These substituents are present in minor proportions, typically 5 to 10 mol. percent. In this chapter the various commercial types and their applications are discussed in sequence, starting with anionic and finishing with cationic resins.

One of the first applications of polyacrylamide was for use in papermaking, where it exhibited outstanding dry strength performance. At that time Swanson (*1*) summarized the situation by saying that many difficultics and inconsistencies were experienced in using beater adhesives, e.g., modified starches. It was hoped that a tailor-made synthetic dry strength polymer would alleviate this situation.

ANIONIC ACRYLAMIDE RESIN

In 1955 an acrylamide based anionic copolymer was introduced to the paper industry for wet end application. This represented the first synthetic polymer specifically designed for dry strength improvement without imparting any wet strength. Unlike many natural resins, acrylamide polymers can be tailor-made with respect to viscosity by molecular weight control and with respect to reactivity by introducing functional groups through copolymerization or hydrolysis.

Acrylamide can be polymerized in solution by using either free-radical catalysts or radiation. Solid acrylamide can also be polymerized by x-ray or gamma radiation. In a typical solution process, a 10% solution of acrylamide in water is heated for two hours at 70 to 80°C in the presence of 0.2% $K_2S_2O_8$ and 10% isopropyl alcohol, with both percentages based on the monomer. The product is a viscous syrup which may be used as such or from which the polymer may be recovered either by drying the solution or by precipitation with acetone or methanol.

Polyacrylamide is an essentially linear nonionic polymer for which the following structure is customarily written:

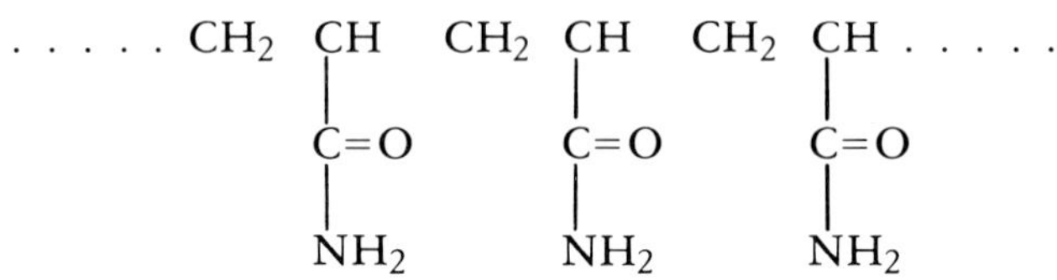

Average molecular weight can be varied from a few thousand to more than 10 million. The polymer is a white powder insoluble in most organic liquids, but it dissolves in water to give clear, viscous solutions. At concentrations at which it is likely to be used, polyacrylamide is compatible with most natural and synthetic gum solutions and latex systems. Under suitable conditions it can act as a flocculant, dispersant, thickener, or adhesive.

One way to make acrylamide polymers substantive to cellulose fibers in dilute aqueous suspension is to introduce ionic groups into the polymer chain. Schiller and Suen (*2*), for example, treated polyacrylamide with alkaline hypochlorite and obtained a polymer containing amino groups. Anionic polymers containing carboxyl groups can be prepared by hydrolysis of some of the amide groups in polyacrylamide or by copolymerization of acrylamide with acrylic acid. Dilute aqueous solutions of these resins or polyelectrolytes behave in a complex way, depending on the pH and on the metal ions present.

The anionic polyacrylamide resin is manufactured in two forms: a dry white powder and a viscous syrup. The powder is easily prepared for use by dissolving in water at concentrations up to 5%. If suitable agitation equipment is used, solutions may be made up to 10% solids. The solution, essentially neutral in pH, possesses excellent storage stability under neutral pH conditions.

The anionic resin is an acrylamide-acrylic acid copolymer (*3*). The molecular weight of the resin, the ratio of the number of carboxyl to amide groups in it, and its other characteristics are such that the resin is an effective slush stock or "beater" additive for paper and paperboard manufacture. For satisfactory results it is used with alum (or an equivalent salt) under controlled pH conditions. Obviously, the resin is best suited for application to rosin sized grades, although unsized papers can be treated if the use of alum is permissible.

Results of Resin Treatment

The resin is effective in improving such physical properties of paper as tensile strength, bursting strength, folding endurance, internal bonding, and resistance to surface rupture (e.g., wax pick), all of which are related to fiber-fiber bonding strength. In contrast with the development of strength by refining, the aforementioned improved strength properties are obtained with very little change in density (bulk), opacity, or porosity. The resin, per se, does not affect the dimensional stability of paper, although indirectly this property may be improved by reducing the degree of beating and refining.

Typical data illustrating the strength improvement realized from the resin are included in Table 6.1 and Figure 6.1. Data in Figure 6.1 are from laboratory work, while the results shown in Table 6.1 are from mill trials.

Table 6.1
Mill Trial Results From Application of Acrylamide/Acrylic Acid Copolymer

	Mill A	Mill B	Mill C	Mill D	Mill E
Paper machine	Fourdrinier	Fourdrinier	Cylinder	Fourdrinier	Fourdrinier
Grade of paper	Offset (5% clay)	Hot drink stock	Bl. manila lined kraft board	Foil liner	Bond
Pulp furnish	Sulfite-soda	Bl. pine-hard-wood kraft	Sulfite liner remainder kraft & wastepaper	Bl. pine hard wood kraft	Sulfite
Resin added,%	0.5	0.5	0.5	0.5	0.4
Retention of resin, %	.....	80	.....	70	.....
Strength Increase, %					
Burst	25	25	24	16	16
Tensile	20	Before increase	15	16	21
Fold	100	in machine speed & furnish change	.....	100	200
Internal bonding	100	.....	.....	.....	.....
Wax pick	13→16	13→16	21→26	14→17	15→17
Sizing increase					
Writing ink	.....	.....	.....	65→100 sec	154→395 sec
Lactic acid	.....	300→650 sec	42→55 sec	.....	before 60%
Hot coffee	.....	30→50 min	.....	.....	cut in rosin size

Table 6.1 — Continued
Mill Trial Results From Application of Acrylamide/Acrylic Acid Copolymer

	Mill A	Mill B	Mill C	Mill D	Mill E
Other improvements	Less filler & fiber in white water	25% increase in production via better drainage, drying & less refining Increased hardwood kraft from 35 to 55% of furnish	Improved bending Cobb (2 min) 0.37→0.25 g/100 sq cm	.....	17% increase in production Hardwood pulp increased from 15 to 35% of furnish Retention of clay up 60% Size usage reduced 60%

	Mill F	Mill G	Mill H	Mill I	Mill J
Paper machine	Fourdrinier	Fourdrinier	Fourdrinier	Fourdrinier	Cylinder
Grade of paper	Fruit wrap	Milk bottle cover stock	Wet strength spinning paper	Ice cream board	White patent coated carrier board
Pulp furnish	Unbl. sulfite-kraft	Bl. pine & hardwood kraft	Unbl. kraft	Bl. pine-hardwood kraft	Sulfite liner, remainder kraft & wastepaper
Resin added, %	0.5	0.5	0.6	0.5	0.5
Retention of resin, %	.....	.....	65	.....	65
Strength Increase, %					
Burst	25	17	15	Specs. met even with an increase in machine speed	14
Tensile	.....	.....	15		12
Fold	.....	.....	.....		.....
Internal bonding	.....	.....	.....	.....	.....
Wax pick	.....	.....	.....	.....	20 →24
Sizing increase					
Writing ink	.....	.....	.....	.....	.....
Lactic acid	.....	.....	.....	.....	.....
Hot coffee	.....	.....	.....	.....	.....
Other improvements	7.5% increase in production due to better drainage & drying Better adhesion to creping dryer	14% increase in production	.....	10% increase in production 50% reduction in rosin size	17% increase in production 8% tear improvement

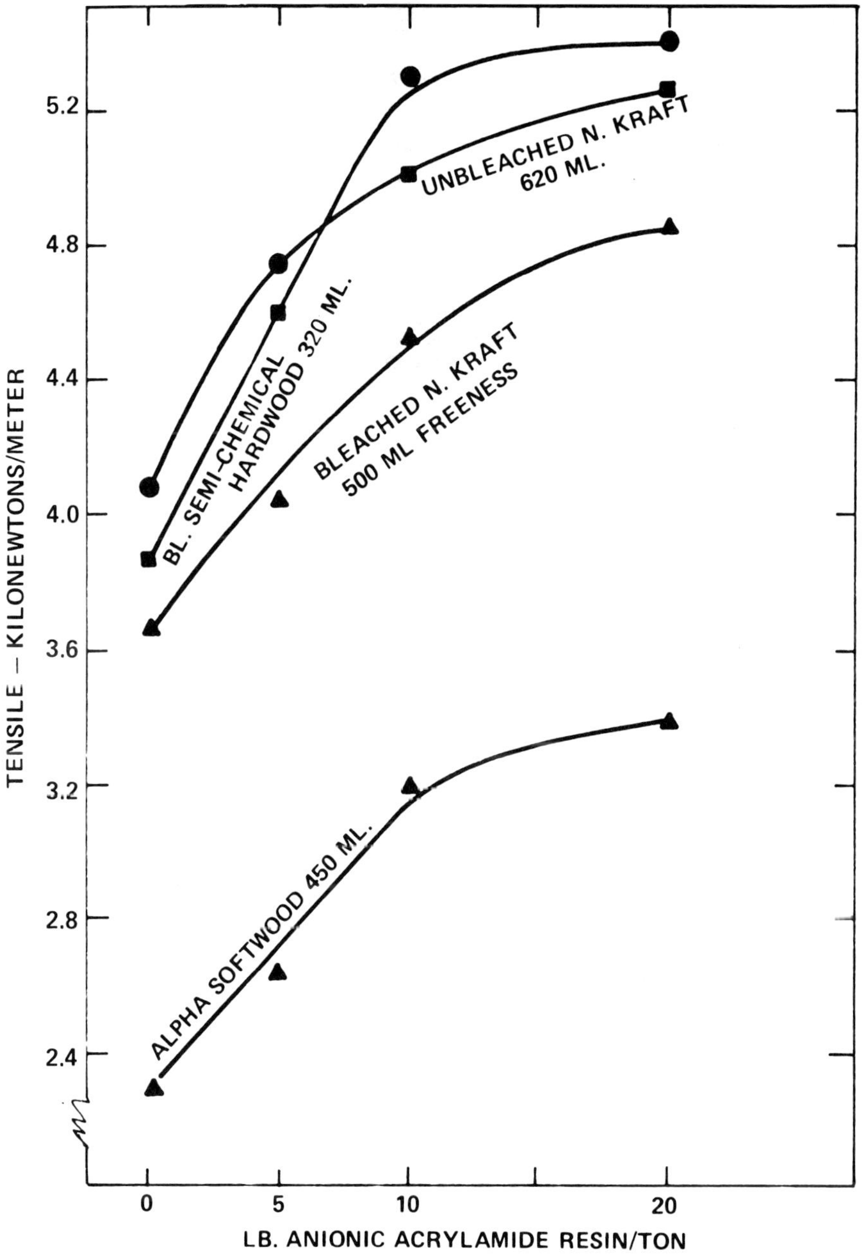

Fig. 6.1. Laboratory data — tensile strength improvement obtained with anionic acrylamide resin for a variety of pulps at different freenesses.

In addition to improving the strength of paper, use of the modified acrylamide polymer frequently provides a number of other valuable features, such as improved retention of fillers, improved rosin sizing efficiency, faster drainage on the wire, and easier drying. Laboratory data illustrating improved rosin sizing are shown in Table 6.2. The ability of the resin to increase substantially production rate via better drainage and easier drying has been observed many times on both fourdrinier and cylinder paper machines. The effect is especially striking when the paper machine is running at peak drying capacity prior to the addition of the resin. Examples for a variety of grades of paper and paperboard are cited in Table 6.1. The increase in production rate made possible by the resin treatment can be further extended by a reduction in refining to free-up the stock. The lower strength resulting from the reduced refining is compensated for by the improvement in strength imparted by the resin.

Table 6.2
Effect of Anionic Acrylamide Resin on Rosin Sizing

Resin treatment (% based on dry fiber)	Ink resistance, sec	20% Lactic acid at 100° F (penescope), sec	% Water absorption (15 min at 73° F)
	Unbleached sulfite pulp		
None	131	215	41.7
0.25	294	575	35.9
	Bleached sulfite pulp		
None	106	55	45.7
0.25	133	190	35.0
	Unbleached kraft pulp		
None	942	1280	35.1
0.25	1464	1850	33.1
	Bleached kraft pulp		
None	202	630	36.6
0.25	255	1065	35.6
	Bleached sulfite-soda pulp blend (60/40)		
None	240	405	28.9
0.25	337	795	27.5

It is well recognized in the paper industry that the effectiveness of any slush stock additive is influenced by the other additives used in the furnish and also by process variables associated with papermaking. The anionic acrylamide dry strength resin described is no exception to this rule. The variables that influence the effectiveness of the dry strength resin are described below. Although they are listed separately, it is clear that a strong interaction exists among some of the variables.

Amount of Resin Used

The resin is effective using as little as 0.1% based on dry fiber. Higher levels of resin treatment, of course, give greater improvements in dry strength. For example, 0.25% resin increased the tensile strength of a moderately beaten softwood-hardwood blend by 18%, while 0.50% resin gave a 32% increase and 1% resin a 47% improvement. The point where the efficiency curve (amount of resin plotted against improvement in dry strength) flattens out is dependent to a large degree upon the type of pulp involved, the amount of refining, and the presence of fillers.

Effect of Pulp Furnish

The application of the resin to a wide variety of pulps is illustrated by the laboratory data in Figure 6.1. The resin is effective with all the commercial grades of pulp tried, with the exception of high (> 50%) groundwood content pulp furnishes. Figure 6.2 shows the effect of 1% resin (based on dry fiber) on the folding endurance of handsheets made from blends of groundwood and unbleached sulfite ranging from 0 to 100% groundwood content.

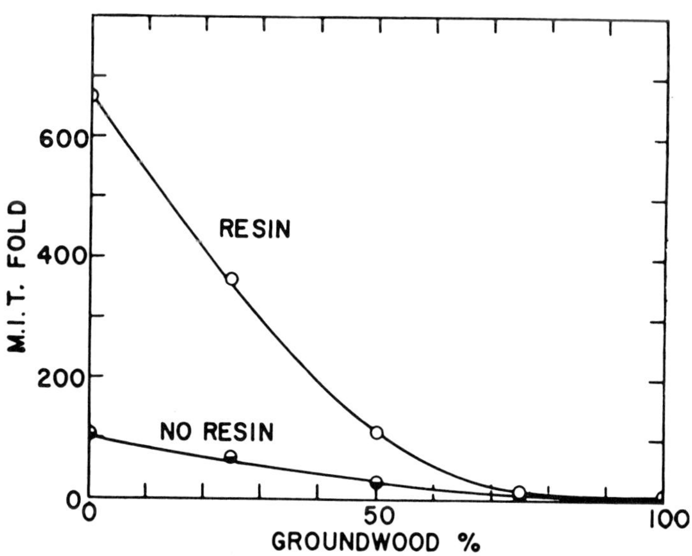

Fig. 6.2 Effectiveness of anionic acrylamide resin in paper containing varying amounts of groundwood.

Alum and Stock pH

For effective performance, the anionic acrylamide resin must be fixed on the fiber by alum, neutralized to the proper extent. The minimum alum [$Al_2(SO_4)_3 \cdot 14H_2O$] requirement is 1.25% based on dry fiber. Greater alum dosages may be required, depending on the amount of rosin size, fillers, or other additives present. Figure 6.3 shows the dry tensile strength imparted by 10 pounds of the acrylamide resin per ton as a function of alum dosage with and without rosin size in a bleached sulfite-hardwood kraft at pH 4.7 (resin added after alum). The importance of controlling the pH of the alum treated stock has been recognized from the beginning of the acrylamide resin development. A typical pH dependence is shown in Figure 6.4.

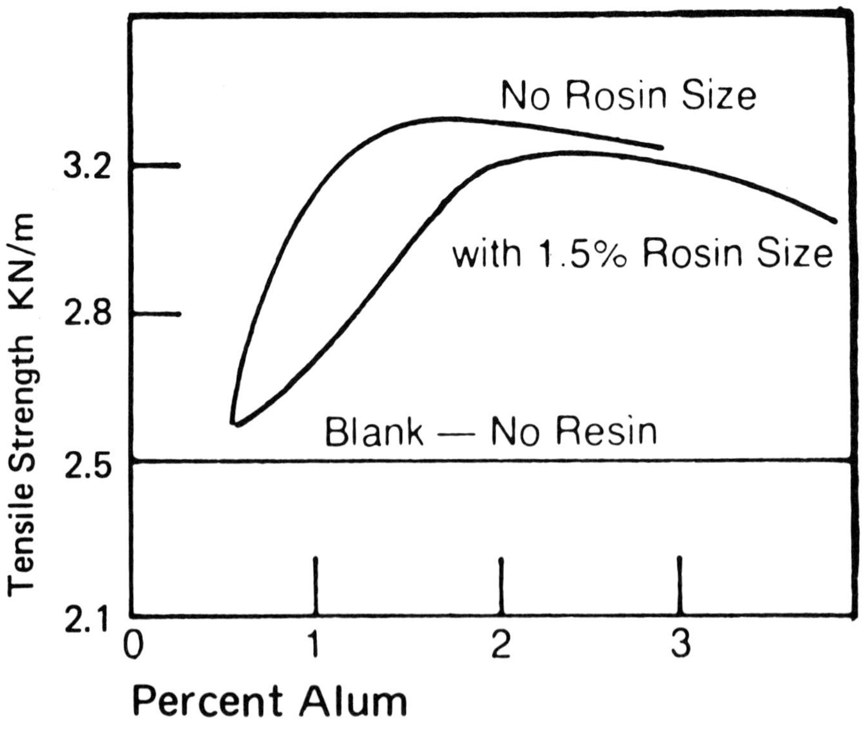

Fig. 6.3. Alum usage versus dry strength imparted by anionic acrylamide resin.

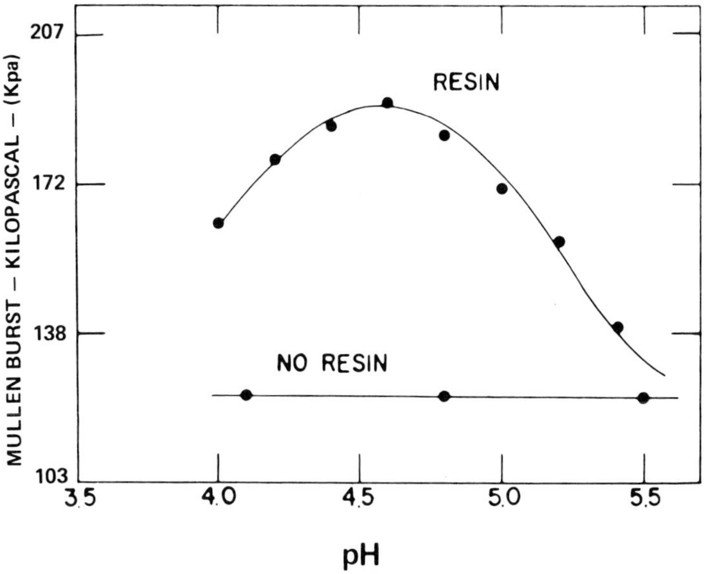

Fig. 6.4. Mullen burst as a function of stock pH.

It was thought that the formation of the most favorable alumina floc was simply a function of pH. This was true as long as the amount of alum and stock consistency remained constant. However, if these two factors are drastically changed, the situation shown in Figure 6.5 occurs. Here, dry strength improvement is plotted against stock pH. In the case of high alum usage and thick stock, the optimum pH occurred at 4.2. At the other extreme, low alum usage and dilute stock, the pH for maximum results was 4.6. Obviously, pH is not the basic factor controlling the formation of the most effective alumina floc.

In Figure 6.6, the data in Figure 6.5 are replotted against the molar ratio of NaOH to alum, instead of pH. It is now apparent what the basic factor is. In Figure 6.6, it should be noted that maximum or near maximum dry strength is realized where the molar ratio of alkali (expressed as NaOH) to aluminum sulfate is between 1:1 and 3:1(*4*).

Degree of Pulp Refining

Figure 6.7 shows the substantial improvement in bursting strength obtained with 0.5% acrylamide resin (based on dry fiber) in a bleached sulfite pulp containing 2.0% alum at stock pH 4.7 over a Canadian Standard Freeness range of 470-35 ml.

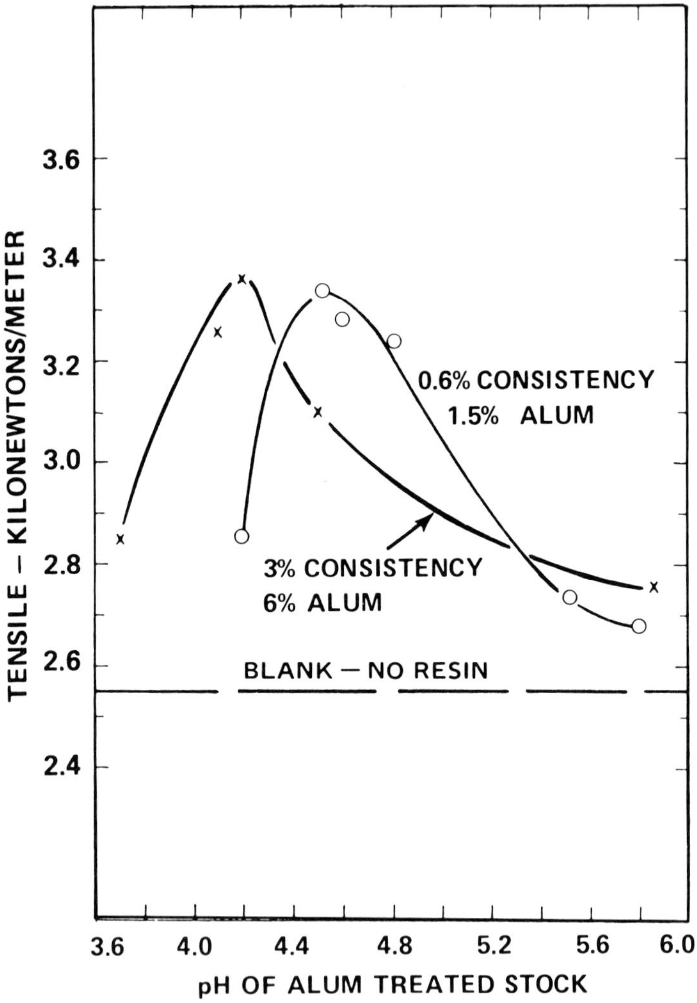

Fig. 6.5. Dry strength versus stock pH; performance of anionic acrylamide resin (0.5% on fiber).

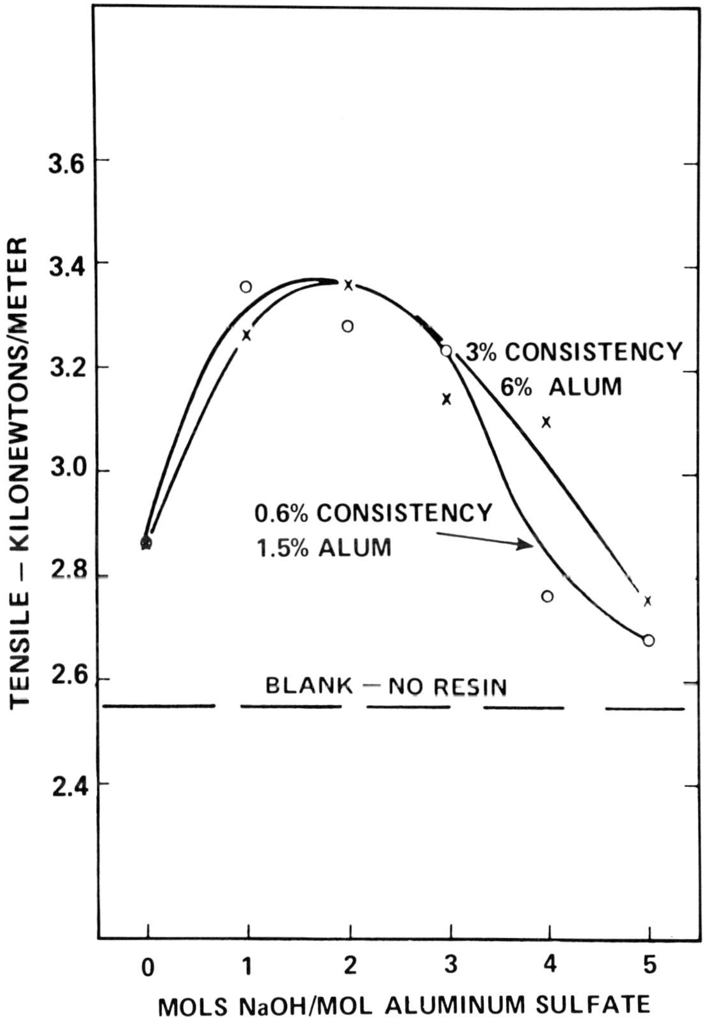

Fig. 6.6. Molar ratio of NaOH/alum versus dry strength.

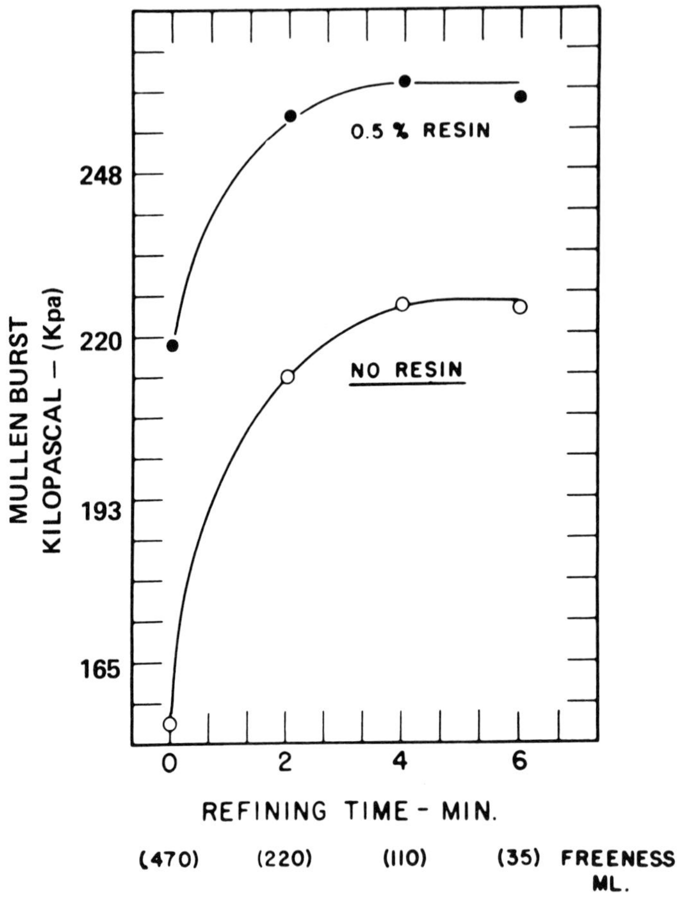

Fig. 6.7. Effect of the degree of pulp refining on dry strength performance of anionic acrylamide resin.

Points of Resin Addition to Stock

In deciding whether to add the resin before or after stock refining, the point of alum addition and the amount of refining are the determining factors. Various sequences of rosin size, alum, and resin additions relative to stock refining are shown in Figure 6.8. A comparison of Line C with either Line A or Line B shows that, if both the resin and alum are added before refining, the effectiveness of the resin is reduced proportionally to the amount of refining. A light action (50 to 75 ml drop in freeness) has little effect, while a more severe refining can seriously impair the performance of the resin. On the other hand, if the resin has not been fixed on fibers by alum, stock treated with resin can undergo severe refining without any loss in dry strength improvement. Equally good results can be obtained by adding the rosin size and alum before refining and by adding the resin afterwards. This sequence has been widely used for paper machine operation with good results.

Although it is quite self-evident, the importance of good mixing of the resin with the stock, particularly thick stock, should be mentioned. Experience has shown that merely introducing a chemical into a pipeline of stock does not necessarily ensure adequate mixing.

Use in Clay Filled Paper

The same general principles that apply to the use of the resin in unfilled paper are also valid for clay filled paper. As previously pointed out, the minimum alum requirement is somewhat higher, 2 to 3% based on dry fiber, rather than 1.25%. Stock pH requirements remain unchanged.

Although the order of addition of clay, resin, and alum is not particularly critical, laboratory data indicate that the greatest improvement in dry strength and filler retention is realized by adding the resin and alum before the clay. However, good results can be obtained even if the resin must be added to clay filled stock; the alum can be added either before or after the resin.

The greatest benefit from the resin treatment of clay filled paper is realized in papers containing up to about 12% clay content. At clay contents higher than this, the improvement in dry strength is diminished.

Use With Wet Strength Resins

Laboratory data confirmed by mill trial results have shown that the modified acrylamide resin can be used to improve the dry strength of paper which is also treated with either melamine or urea wet strength resin. The extent to which the modified polyacrylamide resin can improve the dry strength depends upon the amount of wet strength resin employed. For example, 0.5% of the acrylamide resin based on dry fiber improves the dry strength of paper treated with 1% melamine resin but has little effect on the dry strength of paper treated with 3% melamine resin.

When used in conjunction with the melamine resin acid colloid process, the dry strength resin addition should follow the melamine resin treatment to avoid a drop in wet strength. Maxwell and Reynolds (5) pointed out that multivalent anions such as sulfate or ferricyanide can have an adverse effect on the wet strength if present in the stock prior to the melamine resin addition. An anionic

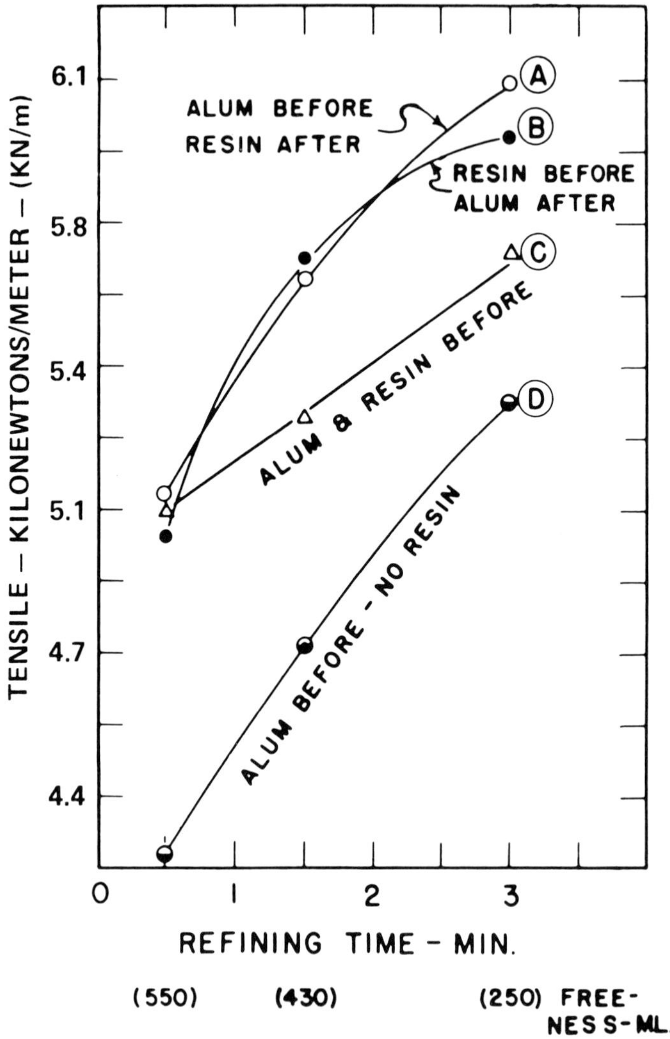

Fig. 6.8. The effect of addition points of alum and anionic acrylamide resin with respect to refiners on dry tensile strength.

polyelectrolyte like the acrylamide polymer behaves in the same way as the inorganic anions. For the cationic-modified urea resins investigated, the order of addition is not critical.

Use With Beater Starch or Gum

In general, the use of beater starch or gum in conjunction with the acrylamide resin has not been found advantageous. Laboratory work has indicated that the use of these materials along with the resin does not materially improve the dry strength over that imparted by the resin. Furthermore, it has been found that starch or gum detracts from the substantial improvement in sizing derived from the resin.

Effect of Dyestuffs

Laboratory experiments have indicated that basic dyestuffs do not affect the performance of the resin. The effect of some of the acid and direct dyes is to increase slightly the amount of alum needed for maximum dry strength. For example, the use of 0.25% CALCOCID[1] alizarine blue (acid) made it necessary to use 2.5% instead of 2% alum, which otherwise would have been adequate. The effect of specific dyes requires individual evaluation.

Fundamental Aspects

From the data presented, it is apparent that in slush stock application both alum and the control of pH are essential for the retention of this anionic acrylamide-acrylic acid polymer by cellulose fibers. As in the case of rosin sizing, positively charged alumina polymer serves to reduce the negative charge of both the cellulose pulp fiber and the resin, so that the resin and alumina polymer may be selectively adsorbed by the fiber from a very dilute resin-pulp mixture. There is, however, at least one important difference between rosin size and this dry strength resin. Under the acidic conditions produced by dilute alum solutions, aluminum rosinates, as well as the undissociated rosin acids of these low molecular weight sizing agents, are very insoluble and form very turbid suspensions. On the other hand, under dilute slush stock conditions the macromolecular anionic dry strength resin and its aluminum salts appear "soluble," showing no visible turbidity. Thus, in the case of rosin size, retention may be due in part to mechanical entrapment among the pulp fibers, as well as to electrostatic bonding. For the dry strength resin, it is believed that retention is due almost entirely to electrostatic and valence bonds.

[1]Trademark American Cyanamid Company

USE OF ANIONIC ACRYLAMIDE POLYMERS WITH CATIONIC POLYMERS

In 1963, the co-use of a cationic resin with anionic acrylamide polymer was introduced. This new way of using the anionic acrylamide resin modified the alum and pH restrictions necessary for the anionic resin. It permitted the application of acrylamide dry strength resin to a wider variety of paper and board grades. According to the patent (6) covering this system, wet strength was not imparted. Davison was issued a patent (7) describing the use of a cationic polyamide wet strength resin with an anionic acrylamide polymer. Davison stated that the cationic wet strength polymer made it unnecessary to use alum with the anionic acrylamide polymer, but, of course, the resulting improvement in dry strength was accompanied by wet strength.

A cationic resin for a non-wet strength system is described in a patent (6) issued to Reynolds as a water soluble non-thermo-setting cationic resin having a molecular weight between about 1,000 and 30,000.

Application of Anionic-Cationic Dry Strength Resin System

The successful use of cationic resin with the anionic acrylamide resin to improve fiber bonding between pH 5.0 and 7.0 depends on several factors, the most important one being the amounts of the two resins. Before a paper machine trial is attempted, laboratory work should be done with the particular stock and white water to determine the optimum ratio of the two resins for maximum results. Total combined resin added to the slush stock is usually in the range of 5 to 15 pounds per ton, based on the dry fiber.

The three-dimensional isometric drawing in Figure 6.9 shows the effect of the anionic/cationic ratio at several levels of total resin. In this particular experiment, 10 pounds of alum per ton at pH 6.0 was used, and the strength test was mullen burst. It is of interest to note that the anionic/cationic (A/C) optimum ratio does not change from 3 to 14 pounds total resin usage.

Some of the factors that influence the ratio of the anionic acrylamide resin and cationic resin to be used are alum, rosin size, pH, pulp purity, fillers, dyes, or other additives. The order and points of addition of the two resins are also important.

Points and Order of Addition

After the amount of the anionic acrylamide resin and the cationic resin to be added have been established in the laboratory, the best points for resin addition should be selected. The two resins should not be added at the same point in the stock system but may be added within a few seconds of each other, provided the

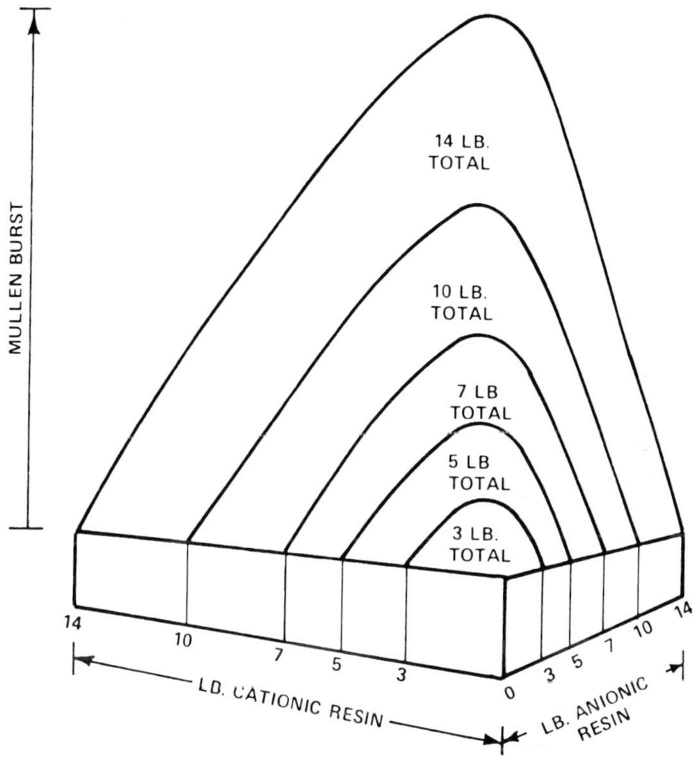

Fig. 6.9. Influence of anionic/cationic resin ratio on dry strength performance.

first resin has been thoroughly diluted and mixed with the stock before the other resin is added. It is good practice to dilute the two resins as much as possible prior to adding to the stock system. The best order of addition is generally the acrylamide resin first, followed by the cationic resin. The stock, after treatment with the cationic resin, should not be subjected to severe shearing action such as is encountered in vortex cleaners. Since no two stock systems are identical, it is impractical to specify exact points of addition that would be applicable to all systems. There are, however, several guidelines for maximum performance and efficiency:

1. Separate points of resin addition should be employed.

2. Preferably, both resins should be added after all stock refining is completed. The acrylamide resin can be added before the final refining if there is no alum present or if, when some alum is present, the stock pH is 5.5 or higher.

3. Where the cationic resin is added late in the stock system, the resin should be highly diluted to facilitate uniform and rapid distribution on the fiber. The actual contact time required may be as little as 3 to 4 seconds.

4. Where rosin size is used, it should be set with alum before any cationic resin is added.

5. The point of alum addition with relation to the acrylamide resin and the cationic resin is quite flexible. Alum may be added before both resins, between the two resins, but not after the cationic resin.

Figure 6.10 is indicative of a system where acrylamide resin is added early in the stock system. This point of addition is feasible only because insufficient alum is present at that point and, therefore, the resin is not deposited on the fiber. In Figure 6.11, the alum is added relatively early in the system in comparison to the resin addition. In the case of modern, large, high speed machines where there is a large volume of stock flow, it is often necessary to have two fan pumps or one fan pump and one booster pump. Figure 6.12 indicates the points of addition which have been successful for optimum efficiency of the resin.

Alum

With modern cationic resins, alum is not essential *unless* rosin size in some form is being used. If so, enough alum to react with the rosin size should be used. Otherwise, the cationic resin will be rendered worthless by the highly anionic rosin size.

pH

The combination of the acrylamide resin and the cationic resin is effective from pH 4.0 to 7.0. It is not necessary to control the pH or total acidity of the tray water within narrow limits to obtain full resin efficiency.

Rosin Size

The type or amount of rosin size employed in the furnish has no adverse effect, per se, on the dry strength performance of the cationic resin and the acrylamide resin combination. Actually, it may be possible to effect an appreciable savings in rosin size as a result of the significant improvement in sizing through the use of these two resins. In some mills, higher sizing tests, e.g., ink and lactic acid resistance, have been obtained at pH 6.0 when using the two resins than are ordinarily obtained at pH 4.5 without them. In addition, higher strength is realized by cutting back on the rosin size, because rosin size itself is detrimental to the strength of paper.

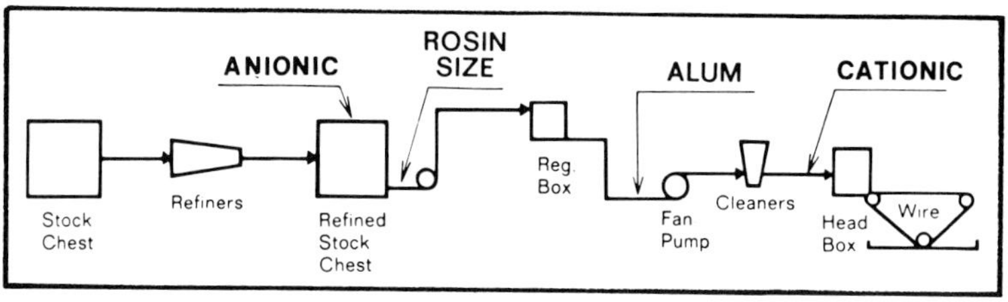

Fig. 6.10. Points of addition of anionic acrylamide resin and cationic resin with relationship to rosin size and alum; acrylamide resin added early in stock system.

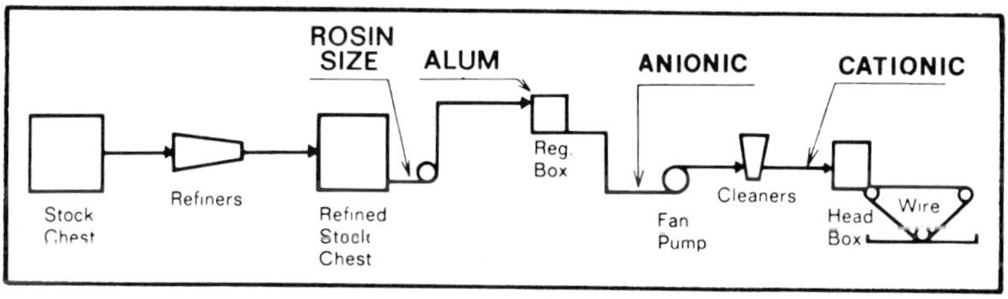

Fig. 6.11. Points of addition of anionic acrylamide resin and cationic resin with relationship to rosin size and alum; alum added relatively early in comparison to resin addition.

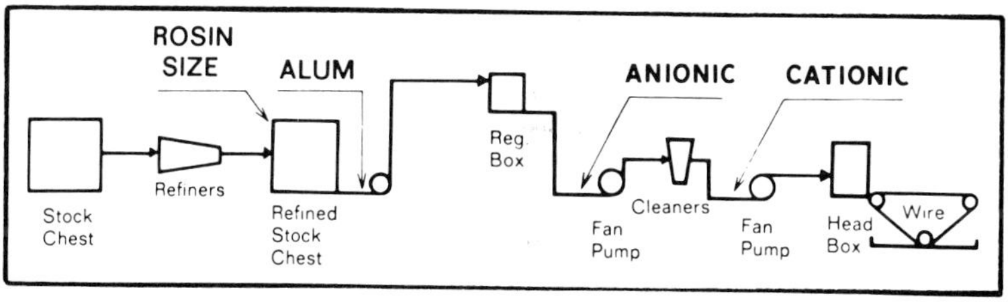

Fig. 6.12. Points of addition of anionic acrylamide resin and cationic resin with relationship to rosin size and alum shown to be successful for optimum efficiency of resin.

Fillers

Whenever cationic resin and acrylamide resin are added to a clay filled stock, the filler retention increases markedly. Unless the ash content is adjusted to its original level, the weakening effect of the additional filler content will minimize and in some instances mask the strength improvement imparted by the resins.

Dyes

Neither direct, acid, nor basic dyes interfere with the application of the cationic/acrylamide resin combination. However, direct or acid dyes should not be added at the same point in the stock system as cationic resin. The intensity of the color is enhanced by the dry strength resin combination. In some cases, the color hue may be affected, but this can usually be compensated for by a modification in the formula.

Commercial Usage

In Table 6.3 are three examples of the successful application of an anionic acrylamide resin with a cationic resin. The resin usage is given as *% total* resin solids based on production. The anionic/cationic (A/C) ratio varied from 7/3 to 5/5 for the examples shown. The same grades made in another mill could have required quite different A/C ratios. The actual optimum ratio is developed on the paper machine by maintaining the total resin usage constant and varying the ratio of anionic to cationic to realize maximum benefits. Increasing one resin by as little as 0.5 lb/ton and decreasing the other by 0.5 lb/ton is common practice in mills successfully using this combination of resins. Drainage and drying rates are more sensitive to such fine tuning than strength tests.

Table 6.3
Commercial Usage of Anionic Acrylamide Polymer with Cationic Polymer

	Mill A	Mill B	Mill C
Paper Machine	Fourdrinier	Fourdrinier	Fourdrinier
Grade	Bleached Food Board	Bleached Food Board	Unbleached Linerboard
Objective	Increase sizing and internal bonds. Higher pH operation.	Increase sizing and strength. Production at higher pH.	Increase production while maintaining sizing specifications.

Table 6.3 — Continued
Commercial Usage of Anionic Acrylamide Polymer with Cationic Polymer

	Mill A	Mill B	Mill C
Total ACCOSTRENGTH® resin Added, %	0.50	0.35	0.30
pH	5.0	5.5	5.5
Strength Increase, % (Compared to untreated paper)			
Mullen	15	30	—
Internal Bond	36	25	—
Sizing Improvement, %			
Lactic Acid Cobb	25	30	—
Water Absorption	—	60	—
Other improvements	After a 22% reduction in rosin size usage. 26% reduction in dryer section steam pressure. 25% reduction in alum usage.	Production increased 10 to 15%.	Production increased 28.5%. Strength and sizing tests maintained at higher production rate.

CATIONIC ACRYLAMIDE DRY STRENGTH RESINS

The three cationic acrylamide dry strength resins available commercially in the U.S. are manufactured by American Cyanamid Company and are described in the sales bulletins (*8-10*) on these resins. Typical physical properties are shown in Table 6.4.

Table 6.4
Typical Physical Properties of Cationic Acrylamide Dry Strength Resins

Trademark	Solids	Color	pH	Viscosity	FDA Status
ACCOSTRENGTH 711	20%	Sl. Yellow	2.5	1000 cps	Approved
ACCOSTRENGTH 410	10%	Clear	4.0	75 cps	Approved
ACCOSTRENGTH 100 UK-A	10%	White Opaque	3.0	1500 cps	Approved

Although the resins are chemically different, the method of application and the benefits to be gained are similar in many respects. Unique features of each resin are discussed separately.

According to the manufacturer (*8-10*), the benefits to be derived from cationic acrylamide dry strength resins include the following:

Higher tensile
Increased burst
Higher fold
Improved internal bond
Improved retention filler and fines
Increased drainage
Improved rosin size performance
Lower steam consumption in dryers

Corollary economic advantages pointed out are use of cheaper fiber, higher productivity, and lower sizing costs.

The method of application common to all three types is as listed.

Amount of Resin and Point of Addition

As little as 0.1% resin based on dry fiber is effective, although 0.2% to 0.5% is ordinarily used to obtain the desired result.

It is recommended that these resins be added in a dilute form after the major portion of the stock refining has been accomplished. Contact time is not critical, although there should be sufficient time to ensure uniform mixing of the resin with the stock. The closer to the headbox the resin is added, the more important are the dilution and the mixing. In most instances, addition to highly diluted stock ($< 0.2\%$ consistency) is not recommended.

Since no two stock systems are identical, it is impractical to specify exact points of addition that would be applicable to all systems. There are, however, a number of guidelines that should be followed to achieve maximum performance and efficiency:

1. Cationic acrylamide resin is always added after the rosin size has been "set" by alum.
2. In stock systems, where vortex cleaners are located between the fan pump and the headbox, dilute cationic polymer resin should be introduced just after these devices if maximum drainage is desirable. Otherwise, addition to the suction side of the fan pump is suggested.
3. The resin should be added to the stock and not to the wire pit water returning to the fan pump.

Rosin Size and Alum

The type or amount of rosin size employed in the furnish does not affect the performance of the cationic resin. Actually, it may be possible to effect an appreciable savings in rosin size as a result of the significant improvement in sizing obtained when the resin is used. In addition, higher strengths are realized by cutting back on the rosin size. In some mills, higher sizing tests, e.g., ink and lactic acid resistance, have been obtained at pH 5.5 when using cationic polymer resin than are ordinarily obtained at pH 4.5 without the acrylamide resin. In the following sections, the influence of pH is discussed for each cationic resin. The necessity of using alum is governed by such factors as rosin, pH, and filler and dyestuff requirements. Alum levels of 0.5 to 1.5% are generally enough, and the use of more than 3.0% alum is not recommended.

Specific information pertaining to each type of cationic resin is shown under separate headings.

ACCOSTRENGTH® 711 Resin

Although applicable to all pulps, this resin has been found to be very effective in bleached grades ranging from sanitary tissue grades to bleached boards. This resin can be used over a broad pH range from 5.0 to 7.5. Since alum is not needed, the resin is applicable to such an absorbent grade as bathroom tissue, where good water absorbency and complete lack of wet strength are required. When alum is used, the resin performs equally well and is thus useful in rosin sized bleached boards.

The full drainage effect of this resin is realized by adding it close to the headbox or forming device. When less drainage is desired, it can be added at the inlet of the fan pump. Extreme dilution of this resin is not required, and it should not be diluted beyond 0.5% concentration.

ACCOSTRENGTH® 410 Resin

This cationic acrylamide resin does not need alum and is effective over an even wider pH range (4.0 to 8.5) than the previous resin. This resin is highly reactive with a limited storage life. The stability of this resin, as shipped, is indicated in Table 6.5. If a high storage temperature or a long storage period is expected, dilution to a lower solids content is recommended in order to prevent gelation.

Table 6.5
Stability of ACCOSTRENGTH® 410 Resin

% Solids	Approx. Days (70°F)	Approx. Days (80°F)
10	29	17
9	40	21
8	72	42
7.5	93	56
7	120	72
6	185	116

According to the manufacturer, this resin can tolerate chlorine and sulfate levels higher than are normally found in most papermaking systems. Concentrations of $SO_3^{=}$ (sulfite ion) in excess of 5 ppm should be avoided.

ACCOSTRENGTH® 100 UK-A Resin

This acrylamide-based cationic resin is especially effective in furnishes containing hydrophobic materials, such as residual cooking liquors and impurities in secondary fiber furnishes.

This resin appears to be especially effective in unbleached kraft linerboard and multi-ply boards containing high percentages of secondary fiber. In these furnishes the resin can strongly influence drainage, drying, and sizing, as well as mullen burst and internal bond.

Unlike the other two cationic acrylamide resins, this resin is effective over a much narrower pH range (5.0-5.5) and needs some alum to attain its full dry strength potential. Generally, 15 to 25 lb of alum/ton suffices.

The manufacturer specifically states that ACCOSTRENGTH 100 UK-A resin must be diluted before it is added to stock. The mixing provided by stock pumps and/or the time available in most stock systems is not adequate to dilute thoroughly the resin as received. Especially designed in-line mixers are available for resin dilution.

Literature Cited

1. J. W. Swanson, *Tappi*, 39(5): 257-67 (1956).
2. A. M. Schiller and T. J. Suen, *Ind. Eng. Chem.*, 48(12): 2132-37 (1956).
3. W. F. Reynolds, L. H. Wilson, W. M. Thomas, N. T. Woodberry, J. C. Barthel, and C. G. Landes, *Tappi*, 40(10): 839-46 (1957).
4. W. F. Reynolds, *Tappi*, 44(2): 177A-179A (1961).
5. C. S. Maxwell and W. F. Reynolds, *Pap. Trade J.*, 126(14): 45 (1948).
6. W. F. Reynolds, U.S. Patent No. 3,332,834 (1967).
7. R. W. Davison, U.S. Patent No. 3,049,469 (1962).
8. American Cyanamid Company, Technical Bulletin PC 105, Wayne, N.J.
9. American Cyanamid Company, Technical Bulletin PC 114A, Wayne, N.J.
10. American Cyanamid Company, ACCOSTRENGTH 100 UK-A Resin Bulletin, Wayne, N.J.

Appendix A

Dry Strength Additives as Practiced in Japan

Morioki Shibuya and Shuji Maruyama

INTRODUCTION

Japan's paper and board production in 1975 amounted to about 13.6 million metric tons (t), the second greatest in the world. In marked contrast to major papermaking countries of North America and northern Europe, however, Japan has very limited wood resources of high quality. This lack of resources is evident from the remarkable increase in the importation of pulpwood since 1965, which accounted for about 40% of the total pulpwood used in 1975, and from the extensive use in papermaking of hardwood pulps, which was as high as 50% in the same year. On the other hand, printers of this oriental civilization have acquired a level of printing technology that can be counted among the world's most advanced in terms of either printing speed or quality. Such advanced technology has in turn called for higher paper quality.

These contradictory circumstances have added ever growing importance to the use of dry strength additives in Japan. The rapidly growing petrochemical industry has supplied paper mills with synthetics at prices more favorably competitive with such natural products as starch or gum and has led to a usage pattern of dry strength additives which is somewhat unique when compared to those overseas.

In the present chapter, the use in Japan of dry strength additives is reviewed by spotlighting some of its special backgrounds. The factors influencing the usage pattern are outlined first and are then followed by a detailed description by type of additive, with stress being laid on such synthetics as polyacrylamide (PAM) and polyvinyl alcohol (PVA). Finally, a brief summary of the dry strength additives used for each type of paper and board is presented.

FACTORS INFLUENCING THE USAGE PATTERN

The changes in the wood resources structure for papermaking in Japan are given in Table A.1 (*1,2,3,4*). Since the supply of softwood, both domestic and imported, could not meet the rapidly increasing demand for pulp, Japanese papermakers, in their efforts to seek new materials, established new technologies for fully utilizing hardwood during the decade that began around

1957. The use of hardwood, characterized by shorter fibers with the resulting weaker paper, continued to increase until 1966, when it attained a share of 55%, with a leveling-off afterwards. The problem of wood resources has been reviewed comprehensively (*5,6,7*).

Table A.1
Changes in the Pulpwood Resources Structure in Japan
Quantity in Thousand m³ with % Breakdown in Parentheses

Year	Softwood	Hardwood	Unknown	Total	Reference
1955	5,764(85.7)	956(14.2)	9(0.1)	6,729(100)	(*1*)
1960	6,695(59.7)	4,522(40.3)	—	11,217(100)	(*1*)
1965	8,034(48.4)	8,579(51.6)	—	16,613(100)	(*2*)
1970	11,693(41.9)	16,208(58.1)	—	27,901(100)	(*3*)
1975	14,561(50.6)	14,225(49.4)	—	28,786(100)	(*4*)

Table A.2 shows a comparison of the domestic and imported wood used by pulp mills (*8,9*). Importation as a whole began to increase around 1965. Although softwood from North America, mainly Douglas fir, still constitutes the greatest part, a remarkable increase has been seen in such wood as eucalypt from Oceanian countries. In addition, 6% of the "domestic" wood represents chips made from lauan wastes. The vessel pick and other problems characteristic of hardwood seem particularly pronounced with such imported wood. The dependence of Japanese pulp mills on the imported eucalypt and other chips will undoubtedly keep increasing for years to come, further adversely affecting the paper quality.

Table A.2
Changes in the Sources of Pulpwood for Papermaking in Japan
Quantity in Thousand m³ with % Breakdown in Parentheses

Year	Domestic wood[a]	Imported wood from U.S.S.R.	North America[c]	Others[d]	Total	Reference
1964	16,601(98.4)	238(1.4)	2 (0.0)	26 (0.2)	16,867(100)	(8)
1967	19,980(92.7)	97(0.5)	1,402 (6.5)	65 (0.3)	21,544(100)	(8)
1970	22,616(81.0)	257(0.9)	4,091(14.7)	937 (3.4)	27,901(100)	(8)
1975	16,996(59.0)	722(2.5)[b]	7,533(26.2)	3,535(12.3)	28,786(100)	(9)

[a]Domestic wood for 1975 consisted of softwood logs (9.4%), hardwood logs (6.3%), softwood chips (29.7%), and hardwood chips (54.6%). Hardwood chips included about one million cubic meters of chips made from waste lauan timbers, corresponding to about 6% of the total domestic wood. Such lauan chips are classified under "Domestic Wood," although all lauan logs are imported.

[b]Chips accounted for 66% of this figure.

[c]The main part of the North American wood consists of Douglas fir and hemlock chips. In 1976, hardwood chips such as alder and oak accounted for about 10% of the total.

[d]In 1975, about 70% of "Others" was eucalypt chips, the rest being waste gum tree timbers, mangrove, and acacia.

Table A.3 shows the wastepaper reclamation rate as the fraction of the wastepaper recycled to the stock out of the total paper consumed (*10*). The 1974 figure for Japan was by far the highest among the major papermaking countries. Such high reclamation rates reflect the poor wood resources situation of that country. A 1975 survey revealed that 0.69 kg of wastepaper was used for producing 1 kg of board, as compared to the corresponding figure of 0.16 for paper (*11*). The inferior strength characteristics of wastepaper could impair the strength of those products in which more wastepaper is used.

Table A.3
Wastepaper Reclamation Rates, 1974 (*10*)

Country	%
Japan	40.8
West Germany	31.7
France	30.6
Sweden	27.8
Finland	25.7
U.S.	21.7
Canada	20.9

During this same period, as rapid economic growth gave rise to the wood shortage and hence the problem of inferior pulpwood quality, Japan experienced remarkable progress in printing technology. In fact, the quality and speed of printing attained as a result of technical innovations required substantial improvements in the printability of paper, including internal and surface bonding strengths. Such requirements, in combination with the unfavorable wood quality circumstance, have placed an extraordinary importance on the necessity of dry strength additives in Japan.

Another important factor influencing the usage pattern of dry strength additives in Japan is that the relative price, and therefore the relative importance of each type of additive, is substantially different from that overseas. For example, Japan depends wholly on imported corn as the raw material for corn starch. Some tapioca and potato starches are imported, as well as all natural gums. Consequently, the market price in Japan of these natural products is substantially higher than it is in the international market.

However, the petrochemical industry offered synthetics at relatively low prices throughout the rapid growth period of 1960 - 1972. Japan is a major supplier of PVA. Even acrylics were more competitive with polysaccharides in Japan than in any other country because of the relatively high cost of the latter. Such an economic background has certainly made Japan one of the countries where the widest use was made of PVA by surface sizing or calender coating and of acrylics by internal addition.

The widespread introduction of on-machine size presses by Japanese paper mills should probably be cited as one of the reasons for the changes in the usage pattern of dry strength additives during these 10 to 15 years. In spite of a rather late beginning, this movement to improve the surface bonding strength of fine printing papers and coating bases grew very rapidly. At the same time there was a switchover from beater sizing with unmodified or pregelatinized starches to surface sizing with oxidized starches.

The constantly expanding needs for reducing organics in paper mill effluents and undernourishing slime-forming bacteria in white water systems have apparently accelerated the replacement of the internally added unmodified starch either by the similarly used acrylics or by oxidized starches or PVA applied at the size press.

STARCH

The volume of starches used either as a dry strength additive or a coating color binder was estimated at approximately 103,000 metric tons (t) in 1976, of which 88% went to paper and 12% to board (*12*).

Historically, the use of starch at the beater was a common practice. This held true until around 1960, when PAM gradually began to replace starch as the main wet end additive for linerboard and sack kraft. At the same time, there was a shift in the use of starch for printing papers from internal to surface sizing because of the widespread use of on-machine size presses, as stated above. For the purpose of internal sizing, the pregelatinized potato starch is most widely used, followed by the unmodified. Its alternative in the internal application is always polyacrylamide, the choice depending on various conditions.

Today, about 70% of papermaking starches are used at the size press, but less than 10% are used internally. The most widely used starch for surface sizing is oxidized corn starch. For printing papers it is sometimes employed in combination with PVA. Good results obtained were explained (*13*) in terms of a synergy in which PVA assisted starch in the strength effect, whereas starch compensated for the high viscosity and readily foaming properties of PVA. The formerly predominant use of starch for board by surface sizing or calender coating appears largely to have been replaced by PVA. Occasionally, unmodified or oxidized starch is said to be sprayed onto board in order to increase the splitting resistance.

Market prices of corn starch, potato starch, and oxidized corn starch for 1976-77 were approximately 85, 160, and 130 yen/kg, respectively.

The cationic starches, although small in the actual amounts now being used, are drawing renewed attention (*14*), as they have proved to be capable of reducing the BOD loading on the waste water treatment systems as well as improving the drainage and retention on wire. Dialdehyde starches (DAS), which improve both wet and dry strengths, have been used since 1964 for facial tissues and for some brands of high quality printing papers by wet end addition or addition at the size press (*15*). No practical use of hydroxyethyl- or hydroxypropyl-starch or of starch phosphates has been reported.

NATURAL GUM

There have been very few reliable sources of information that make it possible to estimate the quantity of natural gums used in papermaking. Unmodified, high viscosity guar gum is the most widely accepted, although new developments such as low viscosity modification and highly cationized derivatives have recently been introduced. However, locust bean gum has not found any measurable acceptance in this field.

The purpose of employing natural gum is to increase paper strengths in general, but sometimes the improvement of bonding strength and splitting resistance of board, sheet formation, and filler or fines retention is particularly intended. Wet end addition is commonplace. Sometimes 3 to 5% of a cationized guar gum is added to the unmodified. The low viscosity modification is used at the size press.

Almost all kinds of paper and board—including printing, glassine, and Indian papers, as well as linerboard and corrugating medium—are considered suitable for the use of natural gum as papermaking additives. Particularly, increasing uses have been found for printing papers containing GP or CGP for which the acrylics have proved less effective.

Perhaps one of the most unfavorable situations for natural gum in Japan is economy, as the country depends upon importation for the entire natural gum supply. In addition, natural gum requires rather complicated dispersing operations, thereby rendering itself less attractive than its main competitor, PAM, which is delivered in solution.

Some technical data are offered in two reference books (*13,16*); a paper (*18*) gives a comparison of an improved natural gum product with polyacrylamide and urea-formaldehyde resin in wet end applications.

POLYACRYLAMIDE

It was around 1960 when PAM as a dry strength additive was introduced into Japan. Although some trials were conducted, the earlier products of U.S. origin were expensive and required an elaborate dissolving operation, so that little immediate use was seen of them. In 1962, however, domestic manufacturers began supplying PAM in the form of aqueous solutions. Since then the price has continued to decline, mainly because of the ever increasing monomer production sustained by the rapid growth of both the petrochemical and the acrylics industries in Japan. The low price coupled with the increasing need for better paper triggered the rapid penetration of PAM into paper mills.

The 1975 shipping of PAM in paper and board applications was estimated at about 14,000 metric tons (t) of solids. Most of this was wet end anionics supplied as 7 to 15% solutions. Alum was used as a retention aid. Also marketed in small quantities were cationized varieties (*17,19,20,21*), as well as products designed for surface sizing (*17,22*) or for improved formation. Strength improvements were generally intended, but improvements in drainage and increases in machine speed were also sought after for board.

A standard anionic resin cost from 450 to 500 yen cif per kg solids in 1977. Each of the big four—Arakawa Chemical, Seiko Chemical, Hamano Industry, and Harima Kasei—operated a number of local polymerization plants sited near big paper mills or in papermaking complexes. The manufacturers jointly adopted an upper limit of 0.2% for the residual monomer contents in their products (*23*).

In 1973, Japan accounted for 47% of the world's acrylamide monomer capacity (*24*) with its annual capacity of 28,000 metric tons (t), as compared to 20,000 metric tons (t) for the U.S., 12,000 metric tons (t) for Holland, and 600 metric tons (t) for France.

During 1974-75, Mitsui Toatsu and Mitsubishi Chemicals switched from the conventional sulfuric acid process for acrylamide monomer to the environmentally more healthy catalytic hydration processes that they had developed. As a result of this technical innovation, the two companies survived as the only manufacturers of the acrylamide monomer in Japan, with a combined annual production capacity of 30,000 metric tons (t) based on solids (*25,26*). Most of the products are now shipped in solutions, either 30 or 50%.

A review on acrylamide paper additives has been published (*27*).

POLYVINYL ALCOHOL

A discussion of polyvinyl alcohol, or PVA, is relevant, since this particular line of vinyl polymers is closely associated with Japan in its historical and economical significance, as seen in Table A.4. In 1972, Japan provided 185,000 metric tons (t), about 60% of the world's production (*28*). Originating in Germany and commercially introduced into the U.S. in 1939 (*29*), PVA became a volume production item in Japan, where it found a unique usage as the partially acetalized PVA named vinylon (*30*). In fact, the Japanese vinylon industry has yielded extensive technology for manufacturing PVA, pulled down its prices, and developed a widespread application technology. An authoritative work on PVA was published in Japan in 1970 (*31*) and another in the U.S. in 1973 (37).

Table A.4
Production Capacity of PVA, 1976[a]

Country	Capacity thousand t/y	% Breakdown
Japan	230	66.8
United States	73	21.2
West Germany	28	8.1
France	5	1.4
Italy	3	0.9
United Kingdom	2	0.6
Spain	2	0.6
Taiwan	1.5	0.4
Total	344	100.0

[a]An estimate by the related Japanese industry

Until about 1960, 70 to 80% of the total PVA produced went to vinylon. Then new applications were developed one after another, and the non-vinylon market became predominant in 1968.

PVA with a DP of between 300 and 2400 in either fully (above 98%) or partially (86 - 90%) hydrolyzed form is commercially available. The mean cif price of PVA in Japan continued to fall until 1972 but then began to rise and reached a 380 - 400 yen/kg level in 1976 - 77.

Tables A.5 and A.6 show the production capacities and shipping records, respectively, of PVA in Japan (*38*).

Table A.5
Production Capacity of Japanese PVA Manufacturers, 1976 (*38*)

Manufacturer	Capacity, thousand t/y
Kuraray	123.6
Nippon Synthetic Chemical	64.8
Denki Kagaku	28.8
Unitika Chemical	24.0
Shin'etsu Chemical	9.0
Total	250.2

Table A.6
Japan's PVA Shipping in 1975 (*38*)

Usage	Shipping, t/y	% Breakdown
Vinylon	40,956	29.4
Export	34,703	24.9
Adhesives	16,412	11.8
Textile sizing	16,035	11.5
Films	13,463	9.7
Papermaking	9,134	6.6
Miscellaneous	8,651	6.2
Total	139,354	100

Since PVA forms stronger films and exhibits stronger adhesion to cellulosic fibers than starch or casein, it has replaced both starch as a surface size and casein as a coating color binder (*13,16,17,32*). The use of PVA in Japanese paper mills continued to grow until 1973, as shown in Table A.7 (*36,38,39,40*). It was estimated that in 1977 approximately 90% was used as surface sizes, 7% as coating color binders, and about 3% as internal sizes for rayon papers.

Table A.7
PVA for Papermaking Uses

Year	t	Reference
1955	200	(*36*)
1960	900	(*36*)
1965	3,720	(*36*)
1970	8,500	(*40*)
1973	11,420	(*38*)
1975	9,134	(*38*)
1977	11,000	(*39*)

PVA has replaced oxidized starches as surface sizes, especially for board and linerboard, where fully hydrolyzed polymers with a DP of about 1700 are generally used (*33*). PVA also possesses such drawbacks as higher viscosities, easy foaming, and deeper penetration, leading to too much pickup and, hence, cost. However, penetration can be suppressed by the well known pretreatment with borax solutions (*13,34,41*). Such properties as penetration, viscosity, and foaming can also be controlled by the combination of PVA with starch. In fact, oxidized corn starch and PVA in mixed dispersions are often used for fine printing papers (*13,16,17,35*). Table A.8 shows a summary of typical size press operations as practiced in Japan (*36*).

Table A.8
Typical Examples of Surface Sizing with PVA (36)

Paper and board grade	PVA concn. %	Pickup %	Machine[a]	Purpose
Kraft liner	2 - 2.5	.4 - .8	W, S	Rub resistance, surface gloss, and ink gloss
Jute liner	1.5 - 4	.3 - 1.2	W	Do.
Uncoated white board	2 - 2.5	.4 - .8	W	Ink gloss, color development, and IGT pick
Food grade carton board	2 - 5	.4 - 2	W, S	Oil resistance
Fine grade printing paper	3 - 6[b]	.5 - 1.5	S	IGT pick and offset printability
Coating bases	2 - 4	.4 - 1	S	Smooth coated surfaces, surface bonding strength, and color inhibition
Single-faced coated paper	3 - 5	.5 - 1.2	S, W, M	Curl prevention (by back sizing)
Oil-proof paper	5 - 10	1.5 - 3	R, M	Oil resistance
Carbon tissue	2 - 4	.4 - 1	S, M	Organic solvent inhibition
Release paper	2 - 10	.4 - 3	R, M	Easy release and inhibition of silicone resin penetration
Glassine paper	2 - 5	.5 - 1.2	S, R	Oil resistance
Banknote paper	7 - 8	1.5 - 2	R	Fold endurance and water resistance

[a]W — Calender; S — Size press; R — Roll coater; M — Miscellaneous
[b]Used with oxidized starches

Rayon paper is a generic name for paper made of rayon fiber alone or in mixture with a certain amount of wood pulp, such as shoji-gami (sliding-door paper), lens polisher, water-proof skin of sanitary napkins, and certain types of facial tissues. Rayon fibers alone or in combination with a certain amount of wood pulp is formed into sheets. PVA is added to the stock at a level of about 5 to 10%. Throughout the press and dryer parts, PVA becomes swollen and softened to function as a binder, rather than as a paper strength additive. PVA is used either as a powder or in the form of staple fibers. The latter shows higher retentions and is preferred to the former in most cases in spite of its higher prices, whereas the former is used in such relatively heavy grades as shoji-gami.

MISCELLANEOUS DRY STRENGTH ADDITIVES

Although limited in quantity, carboxymethyl cellulose (CMC) is used at the size press for improving surface bonding strength and printability of printing papers. It is also used to some extent for white board and linerboard by surface sizing. Surface size effects obtained with CMC have been reported (*16,17,42,43*). Additionally, wet end addition of CMC has been tried after the addition of polyamide-polyamine-epichlorohydrine (PAE) resin to enhance the effect of the latter.

While the PAE resin does not increase the dry strength of paper, melamine-formaldehyde (MF), urea-formaldehyde (UF), and DAS do exhibit some dry strength and are sometimes used in the expectation of this effect. For example, UF resin is added to heavy duty sack paper at a level of 0.1 to 0.2% as solids for improving dry strength. Also, a composition comprising UF and modified starch is marketed for use as an internal dry strength additive.

DRY STRENGTH IMPROVEMENT OF EACH TYPE OF PAPER AND BOARD

For printing papers, PAM and starch are almost exclusively used at the wet end. The choice between them usually depends on the paper machine characteristics or compatibility with other chemicals used. During the decade following the early sixties, the installation of on-machine size presses became quite common, rendering the role of internal sizing less important. Today internal sizing is regarded as nothing but a pretreatment for surface sizing or as a partner to act in concert with surface sizing in improving both paper strength and printability. As the surface size for printing papers, oxidized starch is used normally alone but sometimes together with PVA, as stated earlier. Dialdehyde starches, in combination with oxidized corn starch, PVA, or CMC, were reportedly being used in making non-coated, high grade printing paper (*44*). PAM for surface sizing has not seen much popular usage as yet.

A similar situation exists for the use of dry strength additives in coating bases.

For sack kraft, much PAM and very little starch are used internally. PAM is added at a level of 0.05 to 0.1% solids owf when high quality pulp is employed and 0.3 to 0.8% when lower grade pulp is used. Wet strength products, such as salt packs and garbage sacks, formerly employed UF or MF resins, which have gradually been replaced by PAE since 1973. Surface sizing is seldom applied to wrapping papers, except for bleached kraft papers subject to high grade printing to which starches and/or PVA are applied at the size press.

Household tissues employ PAE resins or DAS (*15*) in the expectation of a certain degree of dry, as well as wet, strength. MF and UF resins, which were once applied, are no longer utilized.

PAM is often added to linerboard and corrugating media at the wet end, the addition levels being strongly dependent on the pulp quality and the required degree of strength improvement. In order to make up for the loss of efficiency sometimes encountered with PAM because of the higher water recycles or the more extensively closed up paper mill, trials are being widely executed to use cationic resins such as PAE or cationized DAS. In some cases, natural gums are preferred. If wet strength is desired, PAE or UF resins are used. For the purpose of improving printability, PVA, starch, or, less frequently, CMC is applied at the size press.

Unmodified or oxidized starches are added to white boards at the wet end or by spraying to achieve high splitting resistance, whereas PVA, starch, and, less commonly, CMC are added at the size press to improve printability.

Dry strength additives are seldom used in newsprints. Because of the extensive use of short fiber wood pulps in Japan, more frequent cleaning of the blanket of the multi-color offset newsprint presses is required. One of the various countermeasures taken is an internal application of guar gum.

CONCLUSION AND ACKNOWLEDGEMENT

In this chapter, the use of dry strength additives as practiced in Japan was briefly described, with particular stress being laid on the singular characteristic of a country with extremely limited wood resources.

The materials contained here are based not only on published references but also on information acquired through correspondence or conversations with a number of leading scholars, engineers, and businessmen in service at universities, associations, and firms in the pulp and paper industry and related fields in Japan to whom the authors' gratitude is due.

Literature Cited

1. Unpublished statistical data of the Forest Div., Japan Paper Association, Tokyo, obtained through S. Ozaki, Oct. 11, 1973.
2. "Yearbook of Pulp and Paper Statistics, 1965," compiled by Research and Statistics Dept., Minister's Secretariat, MITI, Japan, pp. 74-75.
3. "Yearbook of Pulp and Paper Statistics, 1970," compiled by Research and Statistics Dept., Minister's Secretariat, MITI, Japan, pp. 60-61.
4. "Yearbook of Pulp and Paper Statistics, 1975," compiled by Research and Statistics Dept., Minister's Secretariat, MITI, Japan, pp. 48-49.
5. N. Sakurai, *Kamipa Gikyoshi (JAPAN TAPPI)*, 28(10): 468-72 (1974).
6. T. Kayama, *Kami-parupu Gijutsu Taimusu (Pulp and Paper Eng.)*, 17(6): 18-25 (1974).
7. M. Machida, *Kami-parupu Gijutsu Taimusu (Pulp and Paper Eng.)*, 17(10): 11-17 (1974).
8. "Pulpwood Statistics, 1972," compiled by Japan Paper Association, Tokyo, p. 2.
9. "Pulpwood Statistics, 1975," compiled by Japan Paper Association, Tokyo, p. 3.
10. "Koshi Tokei Nempo" (Wastepaper Statistics Yearbook), compiled by Wastepaper Recycle Promotion Center, Tokyo, 1975, p. 184.
11. "Koshi Tokei Nempo" (Wastepaper Statistics Yearbook), compiled by Wastepaper Recycle Promotion Center, Tokyo, 1975, pp. 188-89.
12. *Yakuhin Jumpo (Chemicals Semimonthly)*, 894: 3 (April 10, 1977).
13. H. Hasegawa, "Water-Soluble Polymers," ed. by M. Nakamura, Kagakukogyo-sha, Tokyo, 1973, pp. 170-92.
14. *Yakuhin Jumpo (Chemicals Semimonthly)*, 896: 3 (May 10, 1977).

15. S. Maruyama, *Kami-parupu Gijutsu Taimusu (Pulp and Paper Eng.)*, 10(8): 1-5 (1967).
16. N. Nakamura, "Sizing of Paper," 3rd ed., Kitao Shoseki Boeki, Osaka, 1970, pp. 276-78, 654-55.
17. N. Nakamura, "Sizing of Paper," 3rd ed., Kitao Shoseki Boeki, Osaka, 1970, pp. 287-91, 657-62.
18. K. Kubo, *Kamipa Gikyoshi (JAPAN TAPPI)*, 22(7): 382-88 (1968).
19. *Kagaku Kogyo Nippo (Chem. Ind. Daily)*, May 10, 1977.
20. Arakawa Chemical Co., *Kami-parupu Gijutsu Taimusu (Pulp and Paper Eng.)*, 16(8): 64-68 (1973).
21. *Yakuhin Jumpo (Chemicals Semimonthly)*, 886: 3 (Nov. 10, 1976).
22. *Yakuhin Jumpo (Chemicals Semimonthly)*, 847: 3-6 (Feb. 10, 1975).
23. *Yakuhin Jumpo (Chemicals Semimonthly)*, 855: 7 (June 10, 1975.
24. *Fain Kemikaru (Fine Chemicals Semimonthly)*, 2(9): 22-24 (1973).
25. *Yakuhin Jumpo (Chemicals Semimonthly)*, 870: 3 (Feb. 25, 1976).
26. *Yakuhin Jumpo (Chemicals Semimonthly)*, 849: 3 (March 10, 1975).
27. A Kosugi, *Kobunshi Kako (Polymer Processing)*, 16: 301-306 (1967).
28. *Kami-parupu Gijutsu Taimusu (Pulp and Paper Eng.)*, 16(3): 54-55 (1973).
29. M. Leeds, "Kirk-Othmer Encyclopedia of Chemical Technology," 2nd ed., Interscience, New York, 1970, vol. 21, p. 353.
30. K. Nagano, S. Yamane, and K. Toyoshima, "Poval," Kobunshi Kanko-kai, Kyoto, 1972, pp. 35-54.
31. K. Nagano, S. Yamane, and K. Toyoshima, "Poval," Kobunshi Kanko-kai, Kyoto, 1972.
32. K. Nagano, S. Yamane, and K. Toyoshima, "Poval," Kobunshi Kanko-kai, Kyoto, 1972, pp. 196-209.
33. K. Nagano, S. Yamane, and K. Toyoshima, "Poval," Kobunshi Kanko-kai, Kyoto, 1972, pp. 209-215.
34. K. Nagano, S. Yamane, and K. Toyoshima, "Poval," Kobunshi Kanko-kai, Kyoto, 1972, pp. 210-211.
35. K. Nagano, S. Yamane, and K. Toyoshima, "Poval," Kobunshi Kanko-kai, Kyoto, 1972, p. 211.
36. K. Nagano, S. Yamane, and K. Toyoshima, "Poval," Kobunshi Kanko-kai, Kyoto, 1972, p. 210.
37. C. A. Finch, ed., "Polyvinyl Alcohol, Properties and Applications," John Wiley and Sons, New York, 1973.
38. *Yakuhin Jumpo (Chemicals Semimonthly)*, 888: 5, 12 (Dec. 25, 1976).
39. *Yakuhin Jumpo (Chemicals Semimonthly)*, 930: 3 (Dec. 25, 1978).
40. Unpublished data of Sakubi-Poval Kogyo-kai (The Polyvinyl Acetate and Alcohol Manufacturers' Association), Tokyo, obtained through M. Mano, Feb. 8, 1980.
41. G. Colgan and P. Plante, "Polyvinyl Alcohol," in "Paper Coating Additives," Tappi Monograph Series No. 25, 1963, p. 117.
42. S. Tanaka and T. Kikuchi, *Kamipa Gikyoshi (JAPAN TAPPI)*, 21(4): 197-202 (1967).
43. Daiichi Kogyo Seiyaku Co., *Kami-parupu Gijutsu Taimusu (Pulp and Paper Eng.)*, 11(12): 61-65 (1968); 12(1): 98-103 (1969).
44. The Japan Carlit Co., *Kami-parupu Gijutsu Taimusu (Pulp and Paper Eng.)*, 19(10): 32-33 (1976).

Appendix B

Bibliography

NATURAL GUMS

Aaltio, E., and Jouhikainen, P. The effect of some polysaccharide additions on the properties of paper. *Pap. Puu* 40(11): 561-68 (1958).

Allan, G. G.; Crosby, G. D.; and Sarkanen, K. V. Marine polymers (6); Evaluation of chitosan as a strength additive for alpha-cellulose and unbleached sulfite papers. *TAPPI/CPPA/BPBIF, International Paper Physics Conference Preprints,* 21-25 Sept. 1975, Ellenville, NY, pp. 109-114.

Allan, G. G.; Friedhoff, J. F.; Korpella, M.; Laine, J. E.; and Powell, J. C. Marine polymers (5); Modification of paper with partially deacetylated chitin. In *ACS Symposium Series No. 10*, pp. 172-180, 1975.

Anderson, E. Endosperm mucilages of legumes, occurrence and composition. *Ind. Eng. Chem.* 41(12): 2887-90 (1949).

Anderson, O., and Brunsvik, J. J. Flocculation of streaming fiber suspensions: influence of suspending medium. *Sven. papperstidn.* 64(13): 493-99 (1961).

Attwood, B. W. Effect of shear forces upon fiber suspensions with comments on possible application. *Paper Technology and Industry* 7 (4): 367-70 (1966).

Aucejo, J. G. Application of carrobin gum in the paper industry. *O. Papel* 22(11): 5 (1961).

Barry, J. A. Dilute solution properties of a natural polysaccharide. Ph.D thesis, University of Washington, 1963.

Batlle, J. Vegetable gums in paper manufacture. *O Papel* 19(9): 1-8 (1959).

Beechem, H. A. New water-soluble gums. *Graphic Arts Mon.* 29(2): 98-100 (1957).

Black, W. C. Fiber-bonding agents. *Tappi* 47(1): 201A-204A (1964).

Bolam, F., ed. *Consolidation of the paper web.* 2 vols. London: Technical Section, BPBMA, 1966.

Bolam, F., ed. *Formation and structure of paper.* 2 vols. London: Technical Section, BPBMA, 1962.

Branion, R. The effects of some polysaccharide and petroleum residue binding agents on the properties of rigid insulation board. *Tappi* 47(11): 663-69 (1964).

Bridge, F., and Heaton, J. B. New chemicals in the paper industry and their methods of use. *Chem. Ind.* (17): 533-34 (April 25, 1959).

Canadian Pulp & Paper Industry. Chemical buyers' index. *Can. Pulp Pap. Ind.* 28(7): 23-48b (1975).

Carlson, W. A.; Ziegenfuss, E. M.; and Overton, J. D. Compatibility and manipulation of guar gum. *Food Technol.* 16(10): 50-54 (1962).

Chemical progress. *Pap. Mill News* 81(49): 29-64 (1958).

Chlubek, Z. Upgrading one-side machine-glazed wrapping papers. *Przegl. Papier.* 22(12): 389-90 (1966).

Curran, J. F. Strength chemical utilization. *South. Pulp Pap. Manuf.* 37(4): 16-18 (1974).

Cushing, M. L. An evaluation of combinations of starches and natural gums as papermaking aids. *Tappi* 41(7): 155A-158A (1958).

Cushing, M. L. The use of chemical-starch additives to improve printability of paperboard. *Tappi* 41(9): 166A-167A (1958).

Cushing, M. L. Some recent advances in wet-end additives and implications for future research. *Tappi* 44(3): 191A-194A (1961).

Cushing, M. L., and Schuman, K. R. Fiber attraction and interfiber bonding - the role of polysaccharide additives. *Tappi* 42(12): 1006-16 (1959).

Davison, R. W., Putnam, S. T.; Mashburn, R. T.; and Ware, H. O. A technique for evaluation of internal additives, pulps and refining equipment. *Tappi* 40(7): 499-506 (1957).

Deb, S. K., and Mukherjee, S. N. Molecular weight and dimensions of guar gum from light scattering in solution. *Indian J. Chem.* 1(10): 413-14 (1963).

Dugal, H. S., and Swanson, J. W. Effect of polymer mannan content on the effectiveness of modified guar gum as a beater adhesive. *Tappi* 55(9): 1362-67 (1972).

Dyck, A.W.J. Chemical additives - the papermakers' future. *Am. Pap. Ind.* 52(10): 19-23 (1970).

Erspamer, A. The flocculation and dispersion of papermaking fibers. *Pap. Trade J.* 110(24): 33-8 (1940).

Escolano, J. O., and Semana, J. A. Bag and wrapping papers from kauayan-kiling (bambusa vulgaris). *Philippine Lumberman* 16(5): 36-40 (1970).

Esser, J. A. Guar gains after long effort for farm and market progress. *Chemurgic Digest* 15(12): 4-7 (1956).

Esser, J. A. Guar: its development and uses. *Chemurgic Digest* 17(4): 9-12 (1958).

Ferri, C. M. Important factors to consider when selecting water soluble gums. *Am. Dyest. Rep.* 48(14): 32-34 (1959).

Freedman, H., and Goldstein, A. M. "Chemically beaten" pulp. *Pap. Mill News* 84(28): 11-15 (1961).

Geankoplis, C. J. Preliminary study of guar and locust bean gum in papermaking. *Internal Report C/R* (2489), a private communication. General Mills Chemicals (now Henkel Corp., Minneapolis, MN), 1969.

Goldfrank, H. Guar: its commercial uses and agricultural characteristics. *Chemurgic Digest* 16(2/3): 10-11 (1957).

Goldstein, A. M. Gums as wet-end additives in paper. *Tappi* 40(9): 180A-182A (1957).

Goyal, A. K. Guar gum in paper industry. *Indian Pulp Pap.* 22(4): 237-41 (1967).

Guar gum picks up speed. *Chem. Eng. News* 36(8): 34-6 (1958).

Guha, S.R.D.; Jadhav, A. G.; and Sharma, Y. K. Effect of wet-end additives on bamboo pulp (bleached). *Ippta* 7(3): 235-40 (1970).

Hernadi, S., and Erdelyi, J. Effect of saccharidic additives on the beatability of unbleached softwood kraft pulp. *Papiripar* 15(5): 173-8 (1971).

Hochauz, E. Vegetable gums and their papermaking applications. *Przegl. Papier.* 15(3): 74-7 (1959).

Hochauz, E. Investigations on the use of plant gums in paper manufacture. *Zellst. Pap.* 8(7): 256-61 (1959).

Hochauz, E. The effect of galactomannans on the properties and manufacture of filled papers. *Przegl. Papier.* 15(11/12): 331-5 (1959).

Hochauz, E. How to reduce losses of fibers and fillers. *Przegl. Papier.* 16(8): 243-4 (1960).

Hui, P. A., and Neukom, H. Some properties of galactomannans. *Tappi* 47(1): 39-42 (1964).

The Institute reports on guar. *Pap. Trade J.* 121(25): 38-40 (1945).

Jayme, G., and Murty, V. L. The flocculation of fiber suspensions - its assessment, influence on sheet structure, and regulation by mannogalactan addition. *Das Papier* 16(11): 664-70 (1962).

Junghans, G., and Rapp, A. Some industrial experiences with beater additives of the galactomannan type. *Wochenbl. Papierfabr.* 86(11/12): 526-30 (1958).

Karna, A., and Nordman, L. The use of mannogalactan vegetable gums in the paper industry. *Pap. Puu* 40(2): 45-57 (1958).

Knowles, E. Pulp and paper chemicals now easier in supply, steadier in price. *Pulp Pap. Can.* 77(4): 18-25 (1976).

Kooiman, P. The relation between monosaccharide composition and specific optical rotation of galactomannans from plant seeds. *Carbohydr. Res.* 25: 1-9 (1972).

Kubat, J., and Kubinek, V. Utilization of bleached reed pulp in papermaking. *Pap. Celul.* 7(9): 198, 203-5 (1962).

Kubat, J., and Svaton, M. The addition of mannogalactan gums and of carboxymethylcellulose to paper pulps. *Pap. Celul.* 15(7): 145-51 (1960).

Kubat, J.; Svaton, M.; and Kubinek, V. Mannogalactan additives in the paper industry. *Sb. Vysk. Pr. Odboru Celul. Pap.* (5): 233-68 (1960).

Kubinek, V., and Laifrova, J. The improvement of paper quality by the addition of colloids. *Pap. Celul.* 13(11): 254-56 (1958).

Leech, H. J. An investigation of the reasons for increase in paper strength when locust bean gum is used as a beater adhesive. *Tappi* 37(8): 343-49 (1954).

Lent, D. V., and Nuttall, E. P. Board mills widen usage of mannogalactan gums. *Paperboard Packag.* 47(12): 135-36 (1962).

Loftesness, E. J. Why not investigate continuous preparation of wet-end additives. *Pulp Pap.* 43(12): 135-37 (1969).

Lowe, K. E. Secondary fiber requirements for semi-chemical corrugating medium. *Forming and Recycling of Paper Making Fibers,* Proceedings 15th Annual Pulp & Paper Conference, 13-15 Jan. 1971, Western Michigan University, pp. 84-116.

Lukaszczyk, A. Carubin and guaran, vegetable binders and thickeners. *Tex. Rundsch.* 15(3): 130-4 (1960).

Lukaszczyk, A. The plant derived binders and thickeners carubin & guaran. *Adhäsion* 5(1): 12-15 (1961).

McGuire, T. A.; Hofreiter, B. T.; Mehltretter, C. L.; and Rist, C. E. A method for evaluating deflocculants. *Tappi* 51(2): 94-7 (1968).

Maher, G. G.; Ernst, A. J.; Heath, H. D.; Hofreiter, B. T.; and Rist, C. E. Starch polyethylenimino thiourethane: a wet- and dry-strength agent. *Tappi* 55(9): 1378-84 (1972).

Maj, J., and Szwarcsztajn, E. Flocculation of cellulose fibers in aqueous suspension. II. *Przegl. Papier.* 24(1): 5-9 (1968).

Mastro, R. J. Chemical producers outline supply hot spots; say prices may rise 8-10% a year. *Pap. Trade J.* 160(15): 24-5 (1976).

Moeller, H. W. The development of strength properties in paper with wet-end additives. *Pap. Maker* International Number (London): 110-13 (1965).

Mokhasi, S. G.; Jaspal, N. S.; and Bhargava, R. L. A study of strength development of bamboo unbleached pulps. *Ippta* 8(4): 275-84 (1971).

Moreno, J.A.S. Ceragum: a new agent for the bonding of paper fibers. *O Papel* 19(10): 4 (1959).

Morys, E., and Fibich, J. Experience with the use of mannogalactans. *Pap. Celul.* 22(11): 311-14 (1967).

New "chemical" role for natural gums. *Chem. Week* 85(9): 95-100 (1959).

New dyestuffs and chemical aids for the paper industry. *Tex. Rundsch.* 13(12): 724-26 (1958).

Nissan, A. H. General principles of adhesion, with particular reference to the hydrogen bond. *Formation and structure of paper*, Transactions of the 1961 Oxford Symposium, ed. by Francis Bolam. London: Technical Section BPBMA, 1962, pp. 119-134.

Nittner, E. Carubin and guaran in the paper industry. *Pap. Maker (London)* 141(1): 60-62 (1961).

Nordgren, R. Brief history of the guar development at General Mills, a private communication. General Mills Chemicals, Inc. (now Henkel Corp., Minneapolis, MN), 1972.

Ocken, R. Chemicalization - use of additives in eastern Europe an important factor in strengthening the pulp & paper industry. *Zellst. Pap.* 24(9): 257-59 (1975).

Opie, J. W., and Keen, J. L. Dialdehyde galactomannan gums as wet-end, wet- and dry-strength additives. *Tappi* 47(8): 504-7 (1964).

Paper and paperboard manufacturing (23) formation of a sheet. *Paper* 185(2): 78-81 (1976).

Paper and paperboard manufacturing (17) wet-end additives. *Paper* 184(8): 459-64 (1975).

Paper and paperboard manufacturing (18) wet-end additives. *Paper* 184(9): 525-36 (1975).

Paper, board and converting problems. *World's Pap. Trade Rev.* 150(7): 500-4 (1958).

Paper, board and converting problems. *World's Pap. Trade Rev.* 153(22): 1964, -66, -68 (1960).

Rique, T. Galactomannan gums and "espina corona" (honey locust) gum in the paper industry. *Foll. Tec. For.* (4): 16 pages (Jan. 1960).

Rosenthal, F. Beater additives and their future. *Tappi* 42(11): 158A (1959).

Rowland, B. W. The use of guar in paper manufacture. *South. Pulp Pap. J.* 8(7): 20-4 (1945).

Russo, V. A. Sorption studies of a modified locust bean gum on a bleached sulfite pulp. Ph.D. thesis, Institute of Paper Chemistry, 1959.

Russo, V. A., and Thode, E. F. Sorption studies of a modified locust bean gum on a bleached sulfite pulp. *Tappi* 43(3): 209-18 (1960).

Schmidt, W. Carubin and guaran in the production of paper. *Allg. Pap. Rundsch.* (12): 650-56 (June 20, 1959).

Schwalbe, H. C. Effects of sizing, adhesives and fillers on the formation and consolidation of paper webs. *Consolidation of the paper web*, Transactions of the 1965 Cambridge Symposium, ed. by Francis Bolam. London: Technical Section BPBMA, 1966, pp. 692-740.

Silva, J. E. Increasing the internal strength of paper and board by the use of additives at the wet-end. *ATCP* 7(5): 391-94 (1967).

Skagerberg, W. E. Galactomannan gums and their derivatives in papermaking. *TAPPI Plastics-Paper Conference Preprints,* CA-2006, 30 Oct. - 1 Nov. 1967, Chicago.

Smith, D. I., and Muir, D.T.G. Evaluation of nonfibrous materials for papermaking. *Paper Technology and Industry* 4(1): 35-42 (1963).

Smith, F., and Montgomery, R. *The chemistry of plant gums and mucilages and some related polysaccharides.* ACS Monograph Series No. 141. New York: Reinhold, 1959.

Some newer chemical compounds in papermaking. *Pap. Maker (London)* 151(4): 43-8 (1966); 151(5): 64-9 (1966).

Stacey, M. Industrial uses of carbohydrates - present and future. *Chem. Ind. (London)* (5): 222-26 (1973).

Svaton, M., and Kubat, J. The retention of mannogalactan gums in paper. *Pap. Celul.* 14(10): 223-26 (1959).

Swanson, J. W. The effects of natural beater additives on papermaking fibers. *Tappi* 33(9): 451-62 (1950).

Swanson, J. W. Fiber-to-fiber bonding and beater adhesives. *Tappi* 43(3): 176A-180A (1960).

Swanson, J. W. The science of chemical additives in papermaking. *Tappi* 44(1): 142A-181A (1961).

Swanson, J. W. Chemical additives: new dimensions for papermaking. *Chemical Additives - Application, Theory and Control,* Proceedings 14th Annual Pulp & Paper Conference, Jan. 1970, Western Michigan University, pp. 126-49.

Tookey, H. L., and Clark, T. F. Evaluation of seed galactomannan from cassia species as paper additives. *Tappi* 48(11): 625-26 (1965).

Tookey, H. L.; Ernst, A. J.; Lomar, R. L.; and Wolff, J. A. Evaluation of seed galactomannans from legumes as paper sizes. *Tappi* 44(12): 910-12 (1961).

Tookey, H. L., and Jones, Q. New sources of water-soluble seed gums. *Econ. Bot.* 19(2): 165-74 (1965).

A "universal medicine" for paper ailments. *World's Pap. Trade Rev.* 149(22): 1826-28 (1958).

The use of "Burtonite 7" in papermaking. *Allg. Pap. Rundsch.* (45): 1622, 1624 (Dec. 5, 1966).

Van Derveer, P. D. Conference proceedings review (22nd Plastics-Paper Conf. 1967). *Pulp Pap.* 41(46): 40-45, 54 (1967).

Vegetable deflocculating agents for paper. *Paper* 180(8): 509-10 (1973).

Werbin, S. J. Practical aspects of viscosities of natural gums. *Adv. Chem. Ser.* (25): 5-10 (1960).

Werdouschegg, F. MK. Evaluation of retention and deflocculating agents. Paper presented at the 19th Annual Pulp & Paper Conference, Western Michigan University, 12 March 1975, Kalamazoo, MI.

Westra, H. The flocculation of fiber suspensions. *Papierwereld* 17(10): 289-302 (1963).

Whistler, R. L., and BeMiller, J. N. *Industrial gums.* New York: Academic Press, 1959.

Whistler, R. L., and BeMiller, J. N. *Industrial gums: polysaccharides and their derivatives.* 2d ed. New York and London: Academic Press, 1973.

Whistler, R. L., and Saarnio, J. Galactomannan from soy bean hulls. *J. Am. Chem. Soc.* 79(22): 6055-57 (1957).

Wilfinger, H. Paper auxiliaries & additives, with special reference to environmental protection. *Wochenbl. Papierfabr.* 103(19): 711-15 (1975).

Winchester, A. J. What is a gum? Why is it used? How is it applied? And where? *Pap. Trade J.* 149(51): 38-40 (1965).

Windhaus, G., and Petermann, E. P. Statistics as a method for evaluating paper making auxiliaries. *Wochenbl. Papierfabr.* 98(9): 388-92 (1970).

Wultsch, F., and Schindler, G. Paper additives. *Wochenbl. Papierfabr.* 91(11/12): 541-44 (1963).

Wultsch, F., and Schubert, K. Application and effect of beater additives, with special consideration of plant mucilages of the mannogalactan type. *Wochenbl. Papierfabr.* 85(23): 902-12 (1957).

Wultsch, F., and Schubert, K. Application and effect of beater additives. *Pap. Maker (London)* 139(4): 50-55 (1960).

Yang, C. Effect of freeness, consistency, and additives on formation. *Tappi* 46(10): 169A-172A (1963).

CATIONIC STARCH

Alexander, S. D.; Dobbins, R. J.; and Roscoe, R. W. The influence of dissolved electrolytes in a closed paper mill system. *TAPPI Papermakers Conference Preprints,* 28 Apr. - 1 May 1975, Dallas, TX, pp. 93-99.

Brill, H. C. An evaluation of various beater retention aids for titanium dioxide filler in the presence of chlorinated corn starch. *Tappi* 38(9): 522-26 (1955).

Bristol, K. J. How to reduce BOD levels from size-press starches. *Pulp Pap.* 48(10): 98-99 (1974).

Bristol, K. J., and Brown, G. H. New tool for use at the size press—cationic starch. *Pap. Trade J.* 153(27): 42-44 (1969).

Brown, G. H. Cationic starch improves strength of groundwood papers. *Pap. Trade J.* 153(8): 35-37 (1969).

Casey, J. P. *Pulp and paper,* vol. 2. New York: Interscience, 1960.

CATO® Cationic Starch Newsletter 3(2): (1966) (CATO®: Trademark of National Starch and Chemical Corp.).

Commercially available wet strength and interfiber bonding chemical agents for paper and paperboard. TAPPI C.A. Report No. 33, Aug. 1970.

Cushing, M. L., and Schuman, K. R. Fiber attraction and interfiber bonding—the role of polysaccharide additives. *Tappi* 42(12): 1006-16 (1959).

Davidson, R. L., and Sittig, M. *Water soluble resins*. New York: Reinhold, 1968.

Dugal, H. S., and Swanson, J. W. Effect of polymer mannan content on the effectiveness of modified guar gum as a beater adhesive. *Tappi* 55(9): 1362-67 (1972).

Foster, W. A. Water soluble polymers as flocculants in papermaking. *Polym. Sci. Technol.* 2(8): 3-19 (1973).

Goldstein, A. M. Gums as wet-end additives in paper. *Tappi* 40(9): 180A-182A (1957).

Greif, D. S. Cationic starches as paper coating binders. *Tappi* 43(3): 254-60 (1960).

Han, S. T. Retention of small particles in fiber mats. *Tappi* 47(12): 782-87 (1964).

Han, S. T., and Chang, N. L. Retention of fines in fiber mats. *Tappi* 52(4): 688-93 (1969).

Hernandez, H. R. Cationic starch in high groundwood-content papers. *Tappi* 53(11): 2101-4 (1970).

Herrick, R. Effect of starch and other additives upon white water reuse. *Tappi* 49(11): 79A-81A (1966).

Jarowenko, W. Starch. *Encyclopedia of science and technology,* vol. 12. New York: McGraw Hill, 1970.

Leech, H. J. An investigation of the reasons for increase in paper strength when locust bean gum is used as a beater adhesive. *Tappi* 37(8): 343-49 (1954).

Linke, W. F. Retention and bonding of synthetic dry strength resins. *Tappi* 51(11): 59A-65A (1968).

Margotta, D. G. How coating binders add to BOD. *Pulp Pap.* 48(9): 110 (1974).

Marton, J. Fines and wet end chemistry. *TAPPI Papermakers Conference Preprints,* 3-6 June 1974, Boston, MA, pp. 91-99.

Mazzarella, E. D. Modified and unmodified starches as wet end additives. Paper presented at the TAPPI Plastics-Paper Conference, 14-16 Oct. 1963, Cleveland, OH.

Moeller, H. W. Cationic starch as a wet-end strength additive. *Tappi* 49(5): 211-14 (1966).

Muhonen, J. M., and Williams, D. G. A light probe for measuring flocculation of groundwood pulp. *Tappi* 56(10): 117-19 (1973).

Stephenson, W. D., and Moeller, H. W. Cationic starch at the wet end cuts Combined's costs. *Pulp Pap.* 41(47): 41-43 (1967).

Swanson, J. W. The science of chemical additives in papermaking. *Tappi* 44(1): 142A-181A (1961).

Walkush, J. C., and Williams, D. G. The coagulation of cellulose pulp fibers and fines as a mechanism of retention. *Tappi* 57(1): 112-16 (1974).

Williams, D. G. Minimizing chemical and fines buildup in white water by chemical means. *Tappi* 56(12): 144-47 (1973).

Williams, D. G., and Swanson, J. W. Particle retention in papermaking systems. *Tappi* 49(4): 147-51 (1966).

Wurzburg, O. B., and Mazzarella, E. D. U.S. Patent No. 3,102,064 (Aug. 27, 1963).

ACRYLAMIDE-BASED POLYMERS

Allan, G. G., and Reif, W. M. Fiber surface modifications (IX) stereotopochemistry of ionic bonding in paper. *sven. papperstidn.* 54(18): 563-70 (1971).

American Cyanamid Company. *Aluminum sulfate.* Paper Chemicals Department bulletin. Wayne, NJ.

American Cyanamid Company. *ACCOSTRENGTH®*. Technical Bulletin PC-67. Wayne, NJ.

Bikales, N. M., ed. *Encyclopedia of polymer science and technology,* vol. 1. New York: Interscience, 1964.

Carlsson, G.; Lindstrom, T.; and Soremark, C. The effect of cationic polyacrylamide on some dry strength properties of paper. *sven. papperstidn.* 80(6): 173-77 (1977).

Landes, C. G., and Maxwell, C. S. A study of the melamine resin process for wet strength paper. *Pap. Trade J.* 121(6): 51-60 (1945).

Linke, W. F. Polyacrylamide as a stock additive. *Tappi* 45(4): 326-33 (1962).

Linke, W. F. Retention and bonding of synthetic dry strength resins. *Tappi* 51(11): 59A-65A (1968).

McKague, J. F.; Etter, D. O.; Pilgrim, J. O.; and Griggs, W. H. Practical applications of the electrokinetics of papermaking. *Tappi* 53(1): 80-83 (1970).

Reif, W. M. Utilization of ionic bonds in development of paper strength. *TAPPI Papermakers Conference Proceedings,* 5-8 June 1972, Atlanta, GA, pp. 47-61.

Reynolds, W. F. Some recent research on ACCOSTRENGTH® 86 resin. *Tappi* 44(2): 177A-179A (1961).

Reynolds, W. F. New aspects of internal bonding strength of paper. *Tappi* 57(3): 116-120 (1974).

Reynolds, W. F. Commercially available chemical agents for paper and board manufacture. *TAPPI CA Report No. 60,* Papermaking Additives Committee of TAPPI, 1975.

Reynolds, W. F.; Wilson, L. H.; Thomas, W. M.; Woodberry, N. T.; Barthel J. C.; and Landes, C. G. A modified acrylamide polymer for improving the dry strength of paper. *Tappi* 40(10): 839-46 (1957).

Swift, A. M. Polyacrylamide—a new synthetic water soluble gum. *Tappi* 40(9): 224A-227A (1957).

Tanaka, H., and Senju, R. Preparation of cationic polymers and their applications, part I, Hofmann reaction of polyacrylamide. *Kobunshi Robunshu* 33(6): 309-16 (1976).

Tanaka, H., and Senju, R. Preparation of cationic polymers and their applications, part II. *JAPAN TAPPI* 30(7): 392-400 (1976).

Tanaka, H., and Senju, R. Preparation of cationic polymers and their applications, part III. *JAPAN TAPPI* 30(9): 492-500 (1976).

Tanaka, H., and Senju, R. Preparation of cationic polymers and their applications, part IV. *Mokuzai Gakkaishi* 23(4): 204-8 (1977).

Tanaka, H., and Senju, R. Preparation of cationic polymers and their applications, part VI. *JAPAN TAPPI* 31(5): 270-74 (1977).

Thompson, J. O.; Swanson, J. W.; and Wise, L. E. Hemicelluloses and arabogalactans as beater adhesives. *Tappi* 36(12): 534-41 (1953).

Weiner, J. *Dry strength of paper.* Bibliographic Series No. 271. Appleton, WI: Institute of Paper Chemistry, 1976.

Appendix C
Patents Involving Gums

U.S. PATENTS

Patent No.	Issue Date	Author	Assigned To	Title
3,830,736	08/20/74	Werdouschegg, F. MK	General Mills Chemicals, Inc.	Water clarification
3,770,673	11/06/73	Slagel, R. C. Sinkovitz, G. D.	Calgon	Chitosan graft copolymer for making paper products of improved dry strength
3,709,780	01/09/73	Slagel, R. C. Sinkovitz, G. D.	Calgon	Process and product for making paper products of improved dry strength
3,589,978	06/29/71	Kamal, M. R. Keen, J. L.	General Mills, Inc.	Process of making water-repellent paper using a fatty poly-isocyanate and a cationic gum ether and product therefrom
3,471,362	10/07/69	Kent, S. E.	Hodag Chemical Corporation	Starch and gum additive compositions and use thereof in paper-making processes
3,239,500	03/08/66	Keen, J. L. Ward, W. J. Swanson, R. R. Dunning, H. N.	General Mills, Inc.	Preparation of modified polysaccharides

Patent No.	Issue Date	Author	Assigned To	Title
3,236,832	02/22/66	Opie, J. W. Keen, J. L.	General Mills, Inc.	Method of preparing periodate modified polygalactomannan gums
3,225,028	12/21/65	Nordgren, R.	General Mills, Inc.	Acrolein adducts of polygalactomannans and polyglucomannans and process of preparing same
3,223,699	12/14/65	Schlageter, R.	Etablissement Ena	Process for preparing alkylated polysaccharide binding agents
3,205,125	09/07/65	Opie, J. W. Keen, J. L.	General Mills, Inc.	Forming paper containing periodate-oxidized polygalactomannan gum and paper thereof
3,138,473	06/23/64	Floyd, D. E. Ess, R. J. Keen, J. L. Opie, J. W.	General Mills, Inc.	Compositions and process to increase the wet strength of paper
2,970,063	01/31/61	Jordan, W. A. Skagerberg, W. E.	General Mills, Inc.	Process for making dispersible vegetable gums
2,941,661	06/21/60	Piccard, L. E. Swedish, F., Jr.	Minnesota Mining & Manufacturing Co.	Masking tape

Patent No.	Issue Date	Author	Assigned To	Title
2,887,414	05/19/59	Rosenberg, S. Bandel, D.	American Machine & Foundry Co.	Smoking product and method of manufacture
2,868,664	01/13/59	Goldstein, A. M.	Stein, Hall & Co., Inc.	Dry mannogalactan composition
2,803,558	08/20/57	Fronmuller, G. D.	Commonwealth Color & Chem. Co.	Method of treating adhesive gums
2,775,970	01/01/57	Schoenbaum, A. W.	American Tobacco Co.	Cigarette paper
2,767,167	10/16/56	Opie, J. W. Nordgren, R. Hamilton, R. M.	General Mills, Inc.	Process of reducing the viscosity of gums
2,754,207	07/10/56	Schur, M. O. Levy, R. M.	Ecusta Paper Corp.	Process of improving paper formation
2,730,505	01/10/56	Jordan, W. A.	General Mills, Inc.	Increasing the viscosity of guar sols by reaction with formaldehyde
BRITISH PATENTS				
1,136,842	12/18/68	—	General Mills, Inc.	Gum derivatives

Patent No.	Issue Date	Author	Assigned To	Title
1,084,454	09/20/67	Broatch, J. D. Gartland, W. T. Illingworth, J. W. Prior, P. H.	Reed Paper Group, Ltd.	Improvements in or relating to manufacture of corrugated board
1,056,711	01/25/67	—	Nobel-Bozel, S.A.	A method of stiffening corrugated cardboards
965,865	08/06/64	—	General Mills, Inc.	Improvements in or relating to wet strength additives for fibrous cellulosic materials
944,013	12/11/63	—	Minnesota Mining & Manufacturing Co.	Improvements in paper
879,113	10/04/61	Strating, J. Moes, G. Vogel, W. F.	W. A. Scholten Chem. Fabrieken NV	Derivatives of polysaccharides which are soluble or dispersible in cold or hot water

CANADIAN PATENTS

Patent No.	Issue Date	Author	Assigned To	Title
1,007,385	03/22/77	Werdouschegg, F. MK	General Mills Chemicals, Inc.	Water clarification
867,754	04/06/71	Nordgren, R.	General Mills, Inc.	Quaternary ammonium galactomannan gum derivatives

Patent No.	Issue Date	Author	Assigned To	Title
808,531	03/18/69	Tarr, P. S.	Nalco Chemical Co.	Papermaking process
640,980	05/08/62	Miller, A. Magnuson, A. L.	Kelco Co.	Method of making paper and paper-making composition
589,735	12/29/59	Marrone, J. C. Meincke, H. D.	Howard D. Meincke, Inc.	Composition containing starch and gum for use in the manufacture of paper
567,342	12/09/58	Schur, M. O. Levy, R. M.	Olin Mathieson Chemical Corp.	Process of improving paper formation
566,365	11/18/58	Johnson, S. J. Marra, W. H.	S. D. Warren Co.	Starch strengthened paper
563,236	09/16/58	Anker, C. A.	General Mills, Inc.	Mannan type gums
562,272	08/26/58	Schoenbaum, A. W.	American Tobacco Co.	Cigarette paper
556,985	05/06/58	Jordan, W. A.	General Mills, Inc.	Aqueous sol of guar and formaldehyde
544,553	08/06/57	Opie, J. W. Nordgren, R. Hamilton, R. M.	General Mills, Inc.	Process of reducing the viscosity of mannan type gums

Patent No.	Issue Date	Author	Assigned To	Title
FRENCH PATENTS				
1,356,282	02/17/64	—	Nobel-Bozel, S.A.	Process for imparting rigidity to flutes of corrugated board
1,331,383	05/27/63	—	Etablissement Ena, Fa.	Process for the preparation of a binder, especially for cellulose-derived products
1,295,781	05/02/62	Opie, J. W. Keen, J. L.	General Mills, Inc.	Polygalactomannan gum and its preparation
1,250,143	11/28/60	Piccard, L. E. Swedish, F., Jr.	Minnesota Mining & Manufacturing Co.	Improvements in paper
GERMAN PATENTS				
2,439,026	02/27/75	Berner, K.	W. R. Grace & Co.	Agent for sizing paper as well as process and apparatus for preparing same
2,050,849	05/25/72	Guiseley, K. B. Whitehouse, P. A.	Marine Colloids, Inc.	Process for sulfating gums, particularly carbohydrate gums
1,486,014	04/13/72	Engelskirchen, K. Galinke, J.	Henkel & Cie GmbH	Process for preparing hydroxyalkyl ethers of galactomannans

Patent No.	Issue Date	Author	Assigned To	Title
1,946,471	03/18/71	Auhorn, W. Poschmann, F.	Badische ASF AG	Process for manufacture of groundwood containing printing papers
1,244,146	07/13/67	Engelskirchen, K. Galinke, J.	Henkel & Cie GmbH	Process for purifying water-soluble hydroxyalkyl ethers of galactomannans
		NORWEGIAN PATENT		
134,697	08/14/73	Werdouschegg, F. MK	General Mills Chemicals, Inc.	Color removal and water clarification
		SOVIET UNION PATENT		
125,127	12/29/59	Feldman, R. J. Zemlanskij, Y. Y. Rentov, O. S. Goldovskij, E. H.	—	Method of making leatherboard
		SWEDISH PATENT		
388,410	08/17/73	Werdouschegg, F. MK	General Mills Chemicals, Inc.	Color removal and water clarification

Index

TAPPI PRESS Books

The Bleaching of Pulp, 1979, 1963*, 1953*
The Paper Machine Wet Press Manual, 1979, 1972*
Paper Coating Additives, 1978; 1963*
Preparation of Corrugating Adhesives, 1977
Preparation, Circulation, and Storage of Corrugating Adhesives, 1965*
Physical Chemistry of Pigments in Paper Coating, 1977
Paper Coating Pigments, 1976, 1966*, 1958*
Pigments for Paper Coating, 1948*
Synthetic Binders in Paper Coatings, 1975
Protein Binders in Paper and Paperboard Coating, 1975
Testing of Adhesives, 1974, 1963*
Future Technical Needs and Trends in the Paper Industry, 1973
The Sheet-Forming Process, 1972*
Internal Sizing of Paper and Paperboard, 1971*
The Physics and Chemistry of Wood Pulp Fibers, 1970*
Continuous Pulping Processes, 1970*
Technical Evaluation of Petroleum Waxes, 1969*
Management Science in Planning and Control, 1969*
Chemical Recovery in Alkaline Pulping Processes, 1968*
Deinking of Waste Paper, 1967*, 1956*
Operations Research and the Design of Management Information Systems, 1967*
Wet Strength in Paper and Paperboard, 1965*, 1954*
Pigmented Coating Processes, 1965*
Petroleum Waxes - Characterization, Performance, and Additives, 1963
The Training of Supervisors in Corrugated Box Plants, 1963*
The Influence of Environment and Genetics of Pulpwood Quality. An Annotated Bibliography, 1962
Permeability of Plastic Films and Coated Paper to Gases and Vapors, 1962
Synthetic and Protein Adhesives for Paper Coating, 1961*
Protein and Synthetic Adhesives for Paper Coating, 1952
Mechanical Pulp Manual, 1960
Paper Loading Materials, 1958
Water Technology in the Pulp and Paper Industry, 1957*
Industrial Water for Pulp, Paper and Paperboard Manufacture, 1942*
Starch and Starch Products in Paper Coating, 1957*
Starch for Paper Coating, 1947*
Microbiology of Pulp and Paper, 1955*
Anatomy of Common North American Pulpwood Barks, 1954*
Inspection of Digesters, 1954*
Preparation of Paper Coating Colors, 1954*
Machinery for Paper Coating, 1950*
Nature of the Chemical Components of Wood, 1948*
Resins for the Paper Converter, 1947*
Pulpwood Stands, Procurement and Utilization, 1947
Saving Fourdrinier Wires, 1942*

---

*Out-of-print. Available from: University Microfilms, 300 N. Zeeb Road, Ann Arbor, MI, 48106, or University Microfilms Limited, 18 Bedford Row, London, England, WC 1R 4EJ.